CHALLENGE
UNDERGROUND

Cave Conservation

Caves are fragile in many ways. Their features take hundreds of thousands of years to form. Cave animals such as blind fish are rare, and always live in precarious ecological balance in their underground environments. Cave features and cave life can be destroyed unknowingly by people who enter caves without informing themselves about cave conservation. Great, irreparable damage has been done by people who take stalactites and other flowstone features from caves, and who disturb cave life such as bats, particularly in winter when they are hibernating. Caves are wonderful places for scientific research and recreational adventure, but before you enter a cave, we urge you first to learn about careful caving by contacting the National Speleological Society, Cave Avenue, Huntsville, Alabama 35810.

CHALLENGE UNDERGROUND

Bruce L. Bedford

ZEPHYRUS PRESS, INC. • TEANECK

First published in 1975

Library of Congress Cataloging in Publication Data

Bedford, Bruce L   1942-
    Challenge underground.

    (Speleologia)
    Includes index.
    1. Caves. I. Title.
GB602.B4   796.5'25   75-16464
ISBN 0-914264-16-8

    Published in the United States of America, as a title in
the *Speleologia* series, by Zephyrus Press, Inc., 417
Maitland Avenue, Teaneck, NJ 07666.

Printed in Great Britain
in 11pt Plantin type
by Cox & Wyman Limited
London, Fakenham and Reading

To Grandad

Acknowledgements

For their assistance in the preparation of this book I wish to thank John F. Bridge, Mike Boon, Kim Chapin, Dr P. F. S. Cornelius, Joseph K. Davidson, Dave Gill, Dr William R. Halliday, L. Greer Price, Anthony C. Waltham, John Wilmut, Geoff Workman and many others who find their peace underground.

B.L.B.

Contents

Contents

Illustrations

LINE DRAWINGS

All the line drawings were prepared by Leonard Bedford

Chapter 1

Prologue

The silence. The blackness.

Both are absolute, both slipping back to the borders of nothingness. Both having nothing, absolutely nothing to offer you, and yet – for the moment – they are everything you have. Your senses, attuned to the kaleidoscopic mêlée of the other, utterly distant, world, grapple with this nothingness in an attempt to evoke some tiny positive response for your sensation-starved brain.

Intently, you listen to the silence, almost willing your ears to pick up some minute rustle or click. Surely there must be something out there to disturb the passivity of your ear-drums? Some animate fugitive from the light discreetly scavenging amongst the million-year-old debris for food; some particles of dust finally settling into their lodging places in the black? Nothing. Your eyes are equally unresponsive as you strain them, gazing out. It strikes you that you are not quite sure where you should be looking. Straight ahead, yes – but at what distance? Consciously, perhaps for the first time in your life, you move the muscles of your eyes, aware of the slight jerkiness where you expected oiled smoothness. You hold the eye muscles in one position, but could not even hazard a guess as to whether a sudden burst of light would reveal them focused on some object a hundred yards away . . . or on the hands in front of your face.

Suddenly you remember the old cliché for emphasizing bad visibility, and lift your hands in front of your face. Nothing. Quite aware of the futility, you still make some effort at seeking out the familiar outlines, but the blackness has swallowed them completely. Disbelieving that anything could be lightless, you draw your hands closer to your face, and closer, and yet closer . . . the tiny shock of the palm hitting the end of your nose makes you blink. Still nothing.

These informal experiments continue, until you become aware of something strange: suddenly, you are no longer sure whether you are listening to the silence and watching the blackness, or watching the silence and listening intently to the blackness. Again your senses struggle to form themselves into some kind of familiar order, so that at least the experiment can be conducted with the right instruments in the right order. But at this stage you have become an integral part of the black silence, and it is now impossible to sort one from the other. In you and all around you, it makes no difference which sensation you reach for . . . all is submerged in the black miasma of perpetual night. Here there are no clouds to part, revealing a distant star. Here there are no stars, no chance of moonlight, and certainly no sun.

Now you will turn philosophical, and try to accept the situation of isolation – without being able to test that you are still there to be isolated – in a timeless existence where silence and blackness are kneaded together in an untouchable eternity of black cotton wool.

On the verge of panic, you are saved. Your rump, tired of taking your weight on the cold hard surface, begins to ache.

For once saddle-soreness comes as a welcome relief, and your nerve endings savour the food of feeling, tingling rapturously in discomfort. With the barrier breached, other sensations tumble into your brain: the silence is not absolute after all, for there deep inside your chest you hear the obscure thump-thump-thump-thump of your heart. And freedom from extraneous clatter makes you aware of another new noise, the minute schh-schh-schh-schh of pulsing blood detected by sound-starved ear drums.

You are winning your fight in the dark – but is there still no light? Your hands, still hovering in space in front of your face, press gently on your closed eyelids. Nothing yet, but press again in a slightly different spot and a mosaic of coloured shapes springs into life before you, a galaxy of dull-edged stars superimposed over it.

Now you have torn through the blanket which only a few seconds before seemed to be impregnable. You have won, and are a logical, perceptive decision-making human being once again. In control, you take a decision, and remove the helmet from your head. Your fingers find the control tap on the acetylene lamp clamped to the helmet front, click it round a fraction of a turn, then a pause as the water drips down from the brass tank on to the lumps of inert carbide. Only a second or two to wait, then you hear the slight hiss of acetylene gas escaping through the jet. You press your palm over the whole of the reflector, as instructed in the daylight aeons ago, wait until the gas pools slightly in the polished bowl, then jerk away the palm with a motion that simultaneously revolves the flint wheel. There is a loud echoing bang as the gas ignites, and the flame from the jet settles to a steady blue-white inch-long dart. The sudden flood of soft light pours out into the blackness, driving it back into obscure corners and cracks, and you lift your head to gaze.

Before you, around you, is the cave. Your first cave.

It is with surprise that you see the vast passage in front of you, a surprise at the negation of your previously long-held opinion that caves were small and distinctly squalid holes in the ground. This is a hole in the ground, to be sure, but no trace of squalidness here . . . the passage before you, on its three-hundred-foot stretch before it turns the corner into oblivion, is big enough to take four double-decker buses side by side, with another four on top of them, plus another two on top of those, and the whole thing surmounted by yet another acrobatic bus. The light grey walls soar magnificently to their meeting point, a subterranean cathedral with you as the only worshipper; a rather cold and uncomfortable but completely convinced congregation.

This is your first cave, and you smile at the memory of your earlier fears that claustrophobia would claim you as a jibbering victim within a matter of minutes of entering the cave. You know full well that at this point you are hundreds of feet below the sunbathed mountainside above you, and yet the grandeur of the scene before you makes claustrophobia as unlikely as if you were standing in the centre aisle of whichever above-surface cathedral you chose for comparison with your surroundings. True, the approach to this point was through a much smaller passage – sometimes large enough to wander along with two companions at your side, sometimes demanding that you get down on your stomach and slide through a foot-high horizontal fissure. But even where you had negotiated these low bits – 'Not technically a squeeze . . . too big,' the leader had commented nonchalantly – you had felt no trace of fear or the expected phobia.

This discovery astonishes many people who have just made their first caving trip, and the surprise usually does not occur to them until they are plodding back over the moors to a hot meal. Speak of caving to any ardent surface-lover and he or she will invariably shudder slightly, commenting that they couldn't do it themselves, you know, because they suffer from claustrophobia. So many do say this that if they were all to be believed, something approaching 99 per cent of the world's population should seek immediate treatment for acute fear of confined spaces.

When someone does decide to give the sport a once-and-for-all try, they plunge into the entrance with a mental and real gritting of teeth, tensed and ready for the screaming horrors which they know are going to seize their minds after the first few seconds. In truth, the vast majority of beginners never even consider their rocky confinement for they simply do not have the time. All their attention is on the strange movements and contortions which are demanded of them in making any sort of progress underground. Their mind is completely preoccupied with learning the obvious and the subtle tricks of the trade: the correct angle of the head so that the helmet-mounted light will illuminate that part of the ground you wish to step on instead of lighting some arbitrary spot on the roof or picking out highlights on your boot toecaps; the backward jerk of the arm to keep your haversack full of supplies out of the way in narrow places; the constant ducking of the head to avoid headache-breeding clunks against projections from the roof, even though your cranium is quite safe from severe damage inside the tough glass fibre helmet.

Perhaps strangest of all, the novice has to learn to live with his personal cloud. So high is the humidity in most caves that a deep exhalation – and most breathing is full-chested in this activity – produces an immediate dense cloud of vapour a few inches in front of your face. If this home-made climatic freak catches you out in the process of stepping forward on to some small foothold, the result can be unnerving.

You, back in your own first cave, breathe out. You still cannot restrain a smile at the sudden appearance of your very own cloud. On the surface, such a cloud on a bitterly cold winter day will disperse almost as quickly as it appears, with the slightest trace of wind. And anyway, the light of even a pale sun is enough to make walking no problem. In the cave, you quickly found that it was yet another very definite obstacle to overcome.

Whilst the experienced lads had been only too pleased to give you a flurry of advice on how to light your lamp, tuck your boiler-suit trousers into your socks, and a dozen other things, none had even mentioned this breathing business. The truth is, they did not even think of it, this most basic and automatic part of cave movement. But you had watched them carefully after the first frustrating quarter-of-an-hour, and noted their simple but completely effective solution. From then on you too had pushed your upper lip slightly out so that exhaled breaths clouded well down and out of your line of sight.

'Get plenty of chocolate inside you,' the other three had said when they left you sitting on the bank while they made a short reconnaissance of a side passage. 'We'll be back in twenty minutes . . . so don't budge.'

Obediently munching chocolate, it occurs to you that the very last thought in your mind is to wander off on your own, even though you know about the other party coming in an hour later. The bank you are sitting on is a huge one, rolling gently down for sixty feet or so before coming to the floor of the passage. Surprisingly, the leader had referred to it as a mud bank. It had seemed inappropriate to call this clean and dry yellow-brown silt 'mud', for it is quite unlike the surface variety. But this is almost pure mineral, an ultra-fine sand quite different from the surface conglomerate of earth, water, and assorted stinking animal and plant refuse. Some of the mud in the entrance passage had been as sticky as treacle, but at least it was 'clean' and quite acceptable mud. So much for the defamatory cry of 'Caves? Dirty smelly holes in the ground!'

This condemnation of caves as essentially dirty places seems to stem from the fear which caves evoke in most people, a fear which has nothing to do with claustrophobia but is instead one of more basic roots, a feeling that a cave is somehow 'evil'. Fortunately for our species, this fear seems to be confined to fairly recent times in man's history and was not subscribed to by our hairy prehistoric ancestors – quite the opposite, in fact.

You have finished your first bar of chocolate, and your watch tells you that only twenty minutes have elapsed since you were left on the bank, not the two hours you secretly suspected. And there, right on schedule, you hear the returning team. At first just the scraping of steel toecaps on the rock, then the occasional rattle of a stone kicked, and finally the sound of laughter and voices. Eager to show that you are ready for the next part of the trip, along the big passage into the innermost secrets of the cave system, you stand up and shout.

The bellowed 'HELLO THERE!' has an immediate effect. The other team stops instantly, hushed. Then again the small hubbub of their progress down the side passage. 'HELLO THERE!', even louder, and again the same disquieting silence falling immediately, like your own echoes. Feeling foolish – perhaps this is some kind of joke, an indoctrination, with all three suddenly leaping out to scare you – you keep quiet when the noise starts again after a few minutes. Then you get your first urge for a bit of subterranean humour, and clamber behind a boulder. This indoctrination is going to misfire, with you administering the shock for a change.

After ten minutes of uncomfortable crouching, it occurs to you that something, down in your cave, is up. The sounds coming from the other team have not grown any louder than a certain point; in fact they have faded away to silence three or four times. Whilst you accept that your teammates might go to some slight trouble to spring a joke on you, you doubt that they would spend this long scrabbling backwards and forwards along a constricted passage just for the hell of it. Don't worry, dear confused novice, you have just had your first experience of *Them*.

They haunt every cave explored by men, favouring some – usually active ones with plenty of water – more than others. Unlike the personal cloud, there is no known way of dealing with *Them*, not even experience. *They* most frequently make themselves heard when the caver is tired, but this is not a hard and fast rule. The most important thing about *Them* is that they are never seen, only heard, and then indistinctly. *Their* vocabulary is considerable, and one can distinguish the odd single word in a passage of sufficient audio clarity – never a whole sentence, just the occasional word. *They* must regard caves as sheer paradise, for *They* laugh loud and strong even more than *They* talk. *Their* footwork is extremely clumsy, considering *Their* experience, and *Their* progress along any passage is heralded by the din of scraping boots and kicked rocks.

No, experience will never rid the caver of *Them*, but it will teach him one thing: not to look round the corner at *Them* (for *They* are always just around the corner). The early days of caving are punctuated by the novice running up the next stretch of passage to greet whoever it is shouting hello, eager excursions which grow fewer and fewer as he learns that *They* are always round the next corner, even if he looks round the next corner. Instead, the caver must resign himself to enjoying his caves with companions he can never meet beyond a shouted greeting, and content himself with musing on the tiny drips of water and echo-bouncing curved rock walls which make such things possible.

On your mud bank, you are already learning the stranger lessons of caving, and hold back your shout of greeting until you see the lights of the other three bobbing out of the passage and along the slope towards you.

Another ten minutes of waiting while your companions relax with cigarettes and talk of the difficulties they found in the side passage – 'You did right staying here' – and then down the slope and into the cave.

You will soon come to the streamway, and learn the sheer joy of following underground water along the glistening-walled passage, the exhilaration of even a short scramble down through the spray of the stream into the time- and water-scoured pool beneath. You will climb the underground scree mountain and gaze with honest awe at the huge dome above you, a mammoth sunless sculpture fit to cap St Paul's. You will be feeling the first edges of real tiredness as you crawl alone the final low sandy passages, your eyes following the toescraping boots of the man in front. You will begin to wonder when the turning point will be reached, the point at which the leader will stop for a few minutes and then say, quietly, 'Right – out we go'. You will reach that point, the leader will say the words, but you will never want to leave that cave.

You will have seen the treasure. Yes, it is quite true, we do explore our caves in the hope of finding treasure. Something you half suspected all along. But this is cave treasure, with the self-protecting quality of losing all value the instant it is removed from its cave chest. There will be no diamonds, rubies or emeralds in front of you at that place, but a niche lined with bewitching crystals, guarded by an infinitely delicate grill of pure white rock straws hanging from a glittering crystal roof, and twirling helictites stretching like fingers from the walls. At the very entrance to this tiny wonderland, this deep casket of magic, will hang a blood-red calcite curtain, poised as though ready to fall upon the entrancing stage.

You will leave your cave, your first cave, with every limb aching with weariness, your whole body crying out for sleep. But as you climb into your sleeping bag tonight, your last conscious thoughts – like your dreams – will be of that treasure deep in the earth.

Why do people go caving, preferring to spend their weekend and holiday leisure hours deep underground rather than seeking the sunnier pleasures of the surface? This is an enormously difficult question to answer satisfactorily – not just to the satisfaction of the questioner, but to that of the answerer as well. It would be comparatively easy to sound convincing about the physical demands made by a caving trip, or the get-away-from-it-all aspect of the sport, but any caver presenting these as a complete *raison d'être* for their unusual pursuit is not being honest with himself. He could get just as much physical exercise during a day spent in the controlled environment of a gymnasium, but he chooses to go underground. He could enjoy the same element of escapism by climbing or skydiving, but he chooses to go underground.

The easiest way out at this juncture would be to claim that all the individual attractions of caving are combined in their appeal to any one caver. To some extent this is true, too, but then discussion amongst cavers should reveal a common element of attraction far more frequently than it does.

It seems reasonable to claim that there are enough different attractions in caving for each individual to find one which 'clicks' with him, and that he finds the other aspects provide an appealing background to his prime interest. Vague terms, I know, but then I have never met any caver who has been able to explain satisfactorily – even to others of his ilk – exactly what it is that drives him to follow this sport. This difficulty of providing a really sound explanation for one's interests is not confined to caving, though. Ask an ardent football supporter why he spends a considerable amount of money pursuing his home team about the country, and he would be equally hard put to give you really convincing reasons. I like football, he will say. No other reasons? Well, I like to support my own team. No other reasons? I like the company of the other supporters. No other reasons? There's the drinking afterwards, of course. No other reasons? I do like a good sing-song, it's true . . . And so it goes on, through the joys of rattle shaking, scarf waving, souvenir hunting, getting away from it all – not just one reason, but dozens, perhaps hundreds, mixed inseparably so that in the end all one can say with conviction is, 'I like it'.

Unfortunately for cavers, perhaps more than in other neck-risking sport, they usually find themselves giving excuses for going caving, rather than reasons!

Today, more than 16,000 people in Britain alone follow the sport of caving, and the number is rising rapidly. Even if there is no pat answer to the question of why they choose this sport, we can at least list some of the main attractions.

Adventure. We talk about the spirit of adventure as though it were as scarce a commodity in modern man as telepathy, whereas in truth there is more than a grain of it in every man and woman. For those luckless souls who feel the need for some adventurous outlet away from the drab city existence, caving provides plenty of raw materials for the escape. For a start the underground environment is completely unlike any other, despite the previous comparisons between it and our cathedrals. Essentially alien to the naked apes of the twentieth century, a cave provides one of those rare testing grounds between man and nature where he has the chance to come to terms of understanding without the ability to overcome. Like the climber and his cliff, the caver does not try to beat his cave, but to *know* it, and walking that tightrope of survival takes every bit of skill and courage that he can muster.

So the caver finds all the adventure and challenge he could want, but he has to be peculiarly insensitive not to recognize the beauty to be found underground. A few pages ago I took you on your first caving trip, or part way at least, and explained that the highlight for you would be the finding of the 'treasure trove', not just the physical challenge. This was no exaggeration, for the beauty to be found in caves weaves a spell stronger than any great picture or jewel.

There are two distinct types of visual attraction, one mighty, the other frail. In the former category are the magnificent passages and chambers, huge shafts and thundering waterfalls, all inspired by some age-old architect who had the solid rock of the world at his disposal. These are the sights which make one stop and whistle with astonishment, agog at the whole scale of the thing. But it is the beauty of formations which makes you sit and muse in silence.

Stalactites, stalagmites, helictites, curtains, straws, gour pools, cave pearls . . . inadequate names for conveying the beauty of our underground gems. All formed from the dissolved rock of the limestone which surrounds and protects them, taking their often exquisite colours from mineral deposits above, they are more than sufficient excuse for caving for any man who finds them after his arduous struggle through the bleak passages leading to them. Stalactites can descend through countless centuries until they kiss their companion stalagmites on the floor, then both unite to form massive fluted columns. In complete contrast, and yet still formed by the slow drip dripping of calcite-bearing water, are the inverted forests of fine straws. Although made from the rock itself, these can grow to many feet in length, only a fraction of an inch in diameter, and the slightest breath will cause them to sway gently – a clumsy touch shattering them for ever.

Caves are self-contained, but cavers are essentially groups of people with a common interest, and companionship plays a large part in the appeal of the sport. While it is up to the individual to learn how to negotiate a cave by his own skill and muscle, companions make it a shared experience with all the ramifications of extra enjoyment that this means. And if the caver does get into trouble, he can rely on the others to do everything possible to get him to safety. This independence-cum-interdependence plays a great part in the game, which is why a caver will give rather special emphasis to the word teamwork.

I have written of the exploration of caves, but it must be admitted that the use of the word in its purest sense – to search for something new – is not applicable to most caving trips. Whilst there are many areas of the world in which true exploration of caves is possible, with entrances to whole systems never even looked into, discoveries of new caves in Britain, and even in the vast limestone areas on the Continent, are comparatively rare. More common is the search for new passages leading off from the already known parts of our caves, but the discovery of these is not an every weekend affair.

Most caving trips – I dislike the use of the word expedition other than for implying a full scale exploration, and then usually abroad – are simply for sport. Just as the climber can get complete satisfaction from climbing a route scaled by thousands of others, so the caver is happy to make his way to the end of a suitably difficult cave, or the bottom of a pothole, and then return to the surface. Often the team will have the objective of 'poking around' at a promising point of continuation, or have a look into some small side passage or crawl which may have been overlooked by previous visitors, but will not be disappointed if there is no discovery, or the passage does not 'go'.

Of course these sporting trips can result in finds, sometimes very large ones, but most modern cave discoveries are the result of slow painstaking pushing by diggers (underground or on the surface), divers, climbers, or expert cave survey analysis.

There is a certain amount of confusion amongst the public and the Press as to the difference between a cave and a pothole, as both terms are used in Britain; likewise, a caver and a potholer. Caves and potholes are not entirely different things: both are cavities formed underground, but different in layout. Generally speaking, a cave is formed on a more or less horizontal plane, demanding no more than the personal equipment of the caver to explore it. A pothole follows a more vertical development, and its shafts – usually called pitches – dictate the use of ladders and ropes as well as personal equipment. The definitions are not absolute, as some caves do have pitches, but these are usually few and short enough for the original discoverer to opt for the tag of cave at the Christening.

The difference between cavers and potholers is far less distinct, and lies in a combination of geographical location and, to a lesser extent, personal preference. While most clubs in the North of England call themselves such-and-such potholing club, their southern counterparts prefer this-and-that caving club. The prevalence of potholes in the north and caves in the south is responsible rather than inter-regional competition. As caves and cavers have tended to become generic terms in the sport, for the sake of convenience I shall with a few exceptions use them in this book.

The stage is set, and I hope that this introduction has answered some of the questions which had to be answered before I embark on a wider report of the challenges facing cavers today.

Sport, game, pastime, activity, pursuit – call it what you like, caving has come of age. Those who explore our subterranean world now have the skill and equipment to probe deeper than ever before, farther than ever dreamed of, and their discoveries rank no less than those of the mountains, polar regions, rivers, jungles and deserts above them.

Modern cavers are past paying regard to the ill-informed and savage criticism often levelled at their heads. They alone realize that when every square foot of the world's surface has been plotted and photographed from aircraft and satellites, the caves will continue to hold their secrets. And they know full well that they and only they hold the key to those secrets, that their lamps will cast the very first light on those secrets since the world began.

Chapter 2

Three Beginnings

Our story of caves and caving really has three starting points of significance. Three quite separate beginnings, but followed by developments which interweave to provide the fabric as we can see it today.

Without caves the sport of caving would be rather at a loss for somewhere to happen, so we must first try and unravel the process which leads to the creation of a cave. Then, when we have our caves ready and waiting, man enters the picture, twice. First when he discovered the ready-made protection afforded by a cave against our two oldest enemies, bad weather and hungry predators. That was another beginning. The third happened only recently, long after men had learned to build their own protection where it suited them, rather than where they were fortunate enough to find a cave. And this latter introduction follows almost inescapably from the first motives of fear, for after he is finished with fear – or has at least got it under control – a human being is curious. The men who have returned to the caves in our time are asking how, and why?

How did we get our caves? Where do they come from? These are questions which have intrigued speleologists – scientific studiers of caves – for years, but now at least we do have some idea of the answers, even though the finer technical points will keep the experts studying and arguing happily for many years to come.

Perhaps it would be wise first of all to dismiss a few of the wilder notions which some non-cavers have fondly nurtured as explanations for the formation of caves. True caves have not been dug out by men in some ambitious hope of discovering rare minerals; they are quite natural formations. There are a number of so-called caves in Britain and elsewhere which are the result of mining operations, but these do not correctly deserve the name; it can only be supposed that for the purposes of attracting tourists, the name 'cave' carries more appeal and mystique than plain old 'hole' or 'mine'. There do exist hybrid underground systems, however, where it is not unreasonable to use the name – Speedwell Cavern, a popular tourist cave in Derbyshire, is an example, for in this system miners driving a long tunnel did indeed strike a cave system, and this is an integral part of the visitor's tour today.

Another false belief, but a rather appealing one in a way for all that, is that caves have always been there. Whereas, this might better be applied to mountains, which after all are only pinnacles of rock left by upheavals, or the slow erosion of the rock around them, it cannot be said in the case of a cave – unless, that is, we call on the philosophical argument that the cave *was* there all the time, but it happened to be filled with rock.

With three main exceptions, a cave is an underground cavity formed in a layer of limestone rock by the chemical and mechanical action of water. The three exceptions are rock-shelter types of cave, shallow depressions in a rock face caused by the action of wind and rain; sea caves, formed by the action of tides and waves on sea cliffs of limestone or other rock; and lava caves, the often extensive passages and chambers left inside the cold outer crust of a lava flow while the molten rock inside continues to flow. But these are exceptions to the theme of major cave explorations today for they tend to be limited in extent.

Our concern here is with the cave which offers modern explorers the toughest challenges going, caves which can be many miles long and thousands of feet deep. The formula for the birth and development of such systems contains three basic ingredients: limestone rock, slightly acid water, and plenty of time.

Limestone was formed at the bottom of the huge seas which covered much of our planet millions of years ago. Really it seems to be more of a graveyard than an honest-to-goodness rock, for much of it is composed of the countless billions of minute skeletons left by marine animals and plants of those times, with the tiny gaps left between the skeletons plugged by sand and the even more minute remains of micro-organisms. Subjected to pressure over aeons of time, this oozy burial ground was compacted into limestone. The process will continue as long as there are seas with life left in them, the topmost layer of remains adding that fractionally greater pressure on the bottom-most tissue of long-dead generations. When a caver sees the identifiable outline of one of the larger sea dwellers, in the form of a fossil in the wall of a cave, he remembers the unwitting work of the creature and its ancient companions.

So the layers of limestone were formed – some as long ago as 400 million years. But stuck as they were at the bottom of the sea, they would be of purely academic interest to the caver. Only when the tremendous mountain-building forces of the Earth came into play and lifted the limestone high and dry from the oceans, only then was the stage set.

A solid sheet of limestone would offer little scope to any ambitious water running over it, for the water would have – roughly speaking – nothing to get its 'teeth' into, nowhere to bite in to the lower regions of the rock. Our exposed limestone did not remain solid, though, and contains small fissures, cracks or joints as well as faults where the limestone layers are actually displaced.

These fissures provide the key for the surface water to penetrate the rock, but to do its cave-forming work that water must be slightly acid. When carbon dioxide gas combines with water it makes carbonic acid, and this is an acid which corrodes limestone. But the acid produced by water taking carbon dioxide from the atmosphere is really too weak to be very effective in cave formation – our air contains only 0·03 per cent of the gas. The peaty soils which cover many of the cave-bearing limestone masses have much more carbon dioxide trapped inside them, and water trickling and seeping through these soils can achieve a much higher acid level.

Now that the water is acid, it is what can be called 'aggressive', and it turns its aggressive nature to the limestone fissures. At first only chemical corrosion takes place, enlarging the cracks into ever larger tubes and channels, but when these reach a certain critical size erosion enters the process. This is more of a cutting operation, against corrosion's dissolving process. When the channel is large enough to take small grains of sand and rock, these too play their part in the slow erosion of the limestone.

Eventually the water flow will favour the slightly larger channels, and the enlarging process will be limited to these. Fissures become channels, and channels become passages; where a number of cracks have left a comparatively shattered area, large chambers will be formed. As the chamber reaches a certain size, roof collapses will occur, the resulting rock debris being devoured by the stream.

Cave water can be fickle in its habits, and may suddenly find a layer of softer and more soluble rock through which to pierce its course, leaving the original passages inactive and dead.

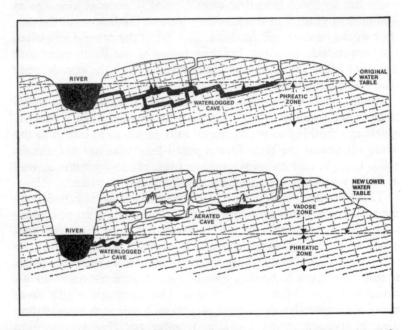

The water table in cave formation. The upper drawing shows the early stage, with a phreatic cave formed in the waterlogged zone below the water table. At the later stage shown in the lower drawing, there is vadose (free-running) water development of caves left by the dropping of the local water table. Phreatic development continues under the water table

Below the water table, the upper limit of that perpetually saturated zone in the ground, the business of building caves continues. But here the water movement is very slow compared with the pell mell rush of water in the larger passages above, only 100 yards a day at times. In this black region of drowned rock, the acidic water is free to eat out a system of vagrant wandering passages, released from the demands of gravity. This is the phreatic zone, with its sharp edged projections and fossil etchings. It is quite distinct, but always susceptible to the smoothing effect of free running water (vadose development) if the water table should fall to a lower level and leave it exposed.

It is impossible to give an exact explanation for the formation of any and all caves; they are all quite individual in character, and while one may display all the signs of sub-water table development another one in the same area will obviously owe its origins to an active underground stream, while yet a third shows signs of a mixture of both processes.

All the variables in the cave-building story – hard or soft limestone, weak or strong acidic water, falling or stationary water tables, great or little rainfall, many or few rock fissures, plus dozens of others – mean that each cave has its own stamp of character. And it is this richness of variety which enables the cave explorer to get complete satisfaction from examining different caves, or even different passages within one large cave. If all caves were pretty much the same, it is doubtful whether he would retain his interest for long.

The illustrations in this book give some idea of the quite different types of passage, chamber and shaft encountered: tall and narrow crevices, low and wide 'bedding planes', circular tubes lifted straight from London's underground, immense chambers with Gothic leanings, plunging shafts with graceful rock-fin adornments.

Three main ingredients were given for the making of caves: limestone, acid water and time. Whilst we can date the approximate age of a sample of limestone by classifying its fossil contents and other means, it is not always so easy to date a cave. We do know, though, that most of the major cave systems in the world today were started from 10,000 to 2,000,000 years ago. Given this length of time, the balanced interplay between water and rock has resulted in the wonderful underground world which we have inherited. Time enough for the slow but unrelenting carving out of such chambers as that in the 250-million-year-old limestone of Carlsbad Cavern, New Mexico, a chamber with an area of 45,000 square yards and a height of 250 feet in parts.

This, then, is the first of our beginnings: that of the caves themselves. The second is that of a most peculiar love/hate relationship, when men first turned their attention to the gaping holes in the cliffs and crags around them, the real cave men, who had to overcome their fear of the dark and unknown not out of curiosity but self-survival.

To the prehistorian delving far, far back into our past, caves are a godsend. While any search in the open plains or hills for clues to man's origins must necessarily be a rather hit or miss affair, caves have acted as storehouses for such evidence. Not all caves contain human remains – comparatively few, in fact – but those that have have offered unique protection from the elements to those tell-tale layers of bones, ashes and artefacts.

Here, in the threshold zone between daylight and darkness, our ancestors found the security to experiment, to develop tools, to fashion utensils, to discover fire. No longer did men have to hide in the open, vulnerable on many sides to attack. In the cave, only the entrance had to be watched – provided, of course, that the family or families had not stumbled upon a cave already inhabited by some animal.

Even with today's sophisticated archaeological techniques of excavation and interpretation, scientists still know remarkably little about the living habits of early man. Radiocarbon and even newer methods of dating artefacts give a fair idea of their age, but much depends on the discoveries of cavers in the future for the piecing together of the cave dweller's everyday life.

Chance has played a great part in the discoveries of some of the most significant subterranean prehistoric finds. In 1856, German quarry workers unearthed some bones in Feldhofer Cave, and a local schoolmaster astutely identified them as those of some prehistoric human being. Not until 24 years after the death of the schoolmaster was he proved right, and the name Neanderthal Man given to the individual who had donated his bit to the cause of science some 50,000 to 70,000 years before.

The skeleton of Cro-Magnon Man, together with stone tools and shell necklaces, was found by a group of French railway workers in 1868. It seems that Cro-Magnon Man was a hunter and food gatherer clad in animal skins, and taking to the caves in winter. But he was something else as well: the first known artist, 30,000 to 40,000 years ago.

The appreciation of the value of art from the Ice Age suffered a serious setback in the last century. A few individual bone engravings and similar items had been found, but then some workmen –

hoping to please the eagerly searching scholars – persuaded a boy to draw some pictures on the wall of a cave. They were quickly recognized as fakes, and for a long time there was an almost blanket refusal to accept cave art as genuine.

Then came the discovery of what is probably the finest painted cave ever found, and again chance played its part, not once but twice.

During a Spanish fox hunt in 1868 one of the hounds fell into a rock fissure. Its master climbed down the crack to rescue the hound, and in doing so uncovered the entrance to the cave of Altamira. There was a gap of seven years, and then the owner of the land, Don Marcelino de Sautuola, began to dig in the cave and found animal bones and flint tools. Some years later he took his twelve-year-old daughter down to see the excavations, but she – no doubt bored with the slow scraping of holes in the floor – wandered into a side passage. It is fortunate that children seem unfettered by the narrow interests of the adult world, for young Maria let her gaze wander not only over the floor but up the walls ... and the ceiling.

Maria dragged her patient father into the side passage to see *her* discovery, and no doubt Don Marcelino tried to look very interested as he flashed his torch over the walls and roof. What his daughter had found utterly astonished him. There in the glow of his torch were the vivid red, black and violet outlines of bulls, wild boars, horses, deer and other animals.

Probably the most famous painted cave in the world is that of Lascaux, in France, and discovery here was also due to an act of canine clumsiness. In fact, it seems that the story of the unearthing of cave art contains more references to lost dogs than deliberate and planned explorations! Lascaux and its spectacular paintings was found by a group of boys in 1940, searching for the inevitable lost dog. The authorities were quick to realize the potential of Lascaux as a tourist attraction, and it was soon opened to the public.

More than 100,000 visitors a year went to Lascaux to stare in wonderment at such richly coloured scenes as the attacking of a masked hunter by a wounded bison. A permanent record of some actual confrontation 20,000 years ago? Or was it of religious signi-ficance, perhaps to give courage to the artist or the tribe? We do not know for certain what prompted such paintings; it was not purely for decorative purposes, for most cave art is hidden in the inner recesses whilst families would undoubtedly have lived nearer the entrance.

The environment of the Lascaux Cave could not stand up to the vast numbers of twentieth-century visitors intent on seeing this most ancient of family albums. In 1960, an ominous green spot of algae appeared on the cave wall, growing rapidly over the following weeks and months. Closer examination revealed other spots of infection. A special commission was set up to fight this menace, and the cave closed to the public for good. Lost in its admiration for the art of our Ice Age ancestors, mankind had unwittingly introduced pollen, sweat, breath and bacteria into the cave of Lascaux, bringing the threat of total destruction to the paintings. Desperate attempts to kill the wall infections without damaging the fragile layers of paint were happily successful. But then there could be no question of the public being readmitted again, and it seemed that their splendours would be witnessed by the very few scientists allowed in and no one else. Fortunately, the landowner of the cave and surrounding area, the Comte de la Rochefoucauld, is taking the gamble of investing £360,000 in a life-size reproduction of the cave and its paintings. The replica, which will be built into the hill next to the actual cave, is being painstakingly fabricated from steel bars and chicken wire covered with cement coloured with the mud from a nearby field. Each contour faithfully reproduces those of the cave walls, and on to the concrete is added the painting itself, executed with a precision accurate to fractions of an inch and in the original raw colouring materials. This is not a project to be hurried, though, and the expected completion date is not until 1981!

Long indeed was the first association between men and caves, and excavations in certain British caves have shown that men occupied them not only up to the Stone Age and Bronze Age, but in Roman times as well. Elsewhere, there are isolated instances of subterranean habitation even today, and the remarkable underground city of Sierra de Guadix still thrives in Spain, with its electrically lit schools, pubs and cinema in grottoes hacked from the rock.

Generally speaking, though, men quit their caves when they

1. Cave entrances may be insignificant fissures a few inches high – or gaping rents in the rock like this, the entrance shaft to Hunt Pot, Yorkshire

2. Fixing tackle in place takes time, so cavers prefer to free-climb where possible, as on the 20-foot Cascade in Giant's Hole, Derbyshire, here being climbed by chimneying technique

developed the ability to build their own dwellings, and their former shelters returned to the darkness and silence once more. Over the years there came about a most unusual change in our attitude to underground places; where our ancestors would – presumably happily – go quite far into a cave's recesses to paint or carve, we developed a fear of such places to the extent that it was a brave man indeed who would dash into the entrance and out again. Tales of demons, dragons, hobgoblins and other assorted malevolent beings were inextricably tied up with 'dark, evil holes'. The more highly developed a race or nation became, the more it seemed to reject with horror those self-same places which nurtured mankind.

Before what can be called the age of exploration, the main visitors to caves were strictly tourists, conducted round amenable and fairly dry systems by enterprising local inhabitants with an eye for the main chance.

The Reverend John Hutton was such a tourist, and his book *A Tour to the Caves*, published in 1780, was one of the first works devoted mainly to underground rather than surface attractions. The English gentry found it fashionable at that time to visit the Lake District, and a few did leave the beaten track to see the caves and potholes near Ingleton and Settle in the Yorkshire Dales; few made any reference in print to what they had seen, though. The reverend gentleman's description of the underground places he was torch-guided through make interesting and often amusing reading these days. The greater the impression made on him by the cave or pot, the more he swung to such descriptive phrases as: 'The profundity seemed vast and horrible from the continued hollow gingling noise, excited by the stones we tumbled into it . . . a subterranean rivulet descends into this terrible hiatus, which caused such a dreadful gloom from the spray it raised up as to make us shrink back with horror, when we could get a peep into the vast abyss.' Quotations from Virgil and Milton stress the horror where necessary.

Then came a growing interest from the scientific side, mainly archaeologists, and with it a little more exploration of non-tourist caves – even if tentatively. When mountaineers first started pursuing their sport as a sport, it was necessary for them to take to the hills festooned with butterfly nets and collecting jars as some kind of scientific justification for an activity which otherwise seemed insane. One wonders whether perhaps a few of those early 'scientific' cave workers found more excitement in their scrambling through unsurveyed caves than in their research?

The birth of serious caving in Britain really dates from the descent of Gaping Ghyll, one of the big Yorkshire systems piercing the flanks of Ingleborough mountain and held to be a sort of granddaddy of potholes. A handful of men had begun the exploration of caves in Yorkshire in the last century, but their sights were set on the big prize: Gaping Ghyll, or GG as most cavers affectionately call it.

One of these men was John Birkbeck of Settle. He knew full well that a descent would be almost impossible in the full force of Fell Beck, the stream which chatters over the moor to plunge into the yawning mouth of Gaping Ghyll. He had a trench dug for over half a mile to divert the stream, then prepared for the descent at the end of a rope. Jerkily the surface team of friends and helpers paid out the rope. When Birkbeck reached a ledge 190 feet down the great shaft, he had to stop his descent. Water was still finding its way into the pothole, and one complete strand of the hauling rope had been cut, on the sharp edges of rock or by a falling stone.

With the formation a few years later of the Yorkshire Ramblers' Club – a club which remained the crack caving organization for many years – plans were made for a really determined assault on the pothole, and members started a tough training campaign in other systems. It is not very difficult to imagine their astonishment and disappointment when, one day in August, 1895, news reached them that their beloved GG had been descended. And to rub salt into the wound, the conqueror had been a Frenchman!

The Father of Caving: that is the title which cavers everywhere bestowed on Edward Alfred Martel, even in the North of England despite the understandable chagrin there. Born in France in the middle of the last century, Martel was completely sold on the idea of scientific cave exploration by the age of twenty. He invented and perfected many of the techniques still used today, and as early at 1890 showed the benefits of portable telephones in the descent of big systems.

Martel heard of the challenge presented by Gaping Ghyll, and gave it pride of place in his itinerary for a tour of English and Irish caves in the summer of 1895.

On Ingleborough, another trench was dug, but again much of Fell Beck's water trickled past the diversion to meet with a roar down the main shaft. Meticulously, Martel prepared his lifelines and ladder. He was aware of the advantages of ladder climbing over a single rope descent, and regularly used the old-fashioned rope ladder with the aid of vigorous tugs from the surface lifelining party. Checking that his crude telephone was still operational, Martel began his descent.

Like Birkbeck, he landed on the ledge at 190 feet. But this time, despite a thorough drenching and buffeting from the stream which thundered past him, the climb down was continued by the Frenchman.

Once past the ledge, the main shaft of Gaping Ghyll opens up into the biggest underground chamber in Britain, nearly 500 feet long. Martel found that the ladder was now hanging free, and he was unable to prevent it swinging, a pendulum motion which carried him into the full force of the stream only to be bounced off again and again.

Tests with a plumbline beforehand had made Martel fairly sure of the shaft's depth, but not positive. With arms now growing numb in the cold and wet of the climb, he suddenly felt the lifeline round his waist grow tight. The ladder was sufficiently long to reach the bottom, but not so the lifeline. He shouted into the telephone for extra rope to be tied on by the surface supporters, but the din of crashing water drowned any reply. Not until the lifeline went slack did he know that his message had got through, and he could continue the descent.

Martel's feet reached for the next rung, but instead found the stones and gravel which cover the floor of GG's main chamber. He was down, and fully deserved the satisfaction of gazing up the 365 feet of ladder which linked him with the surface, its rungs kissed by the highest underground waterfall in England – twice the height of Niagara Falls.

After a brief survey of the chamber, Martel began his climb back to daylight. The vulnerable telephone link was out of action, and rope signals had to be used. The drama of the situation was already high, but when the knotted lifeline jammed in a crack there must have been one or two onlookers who thought the brave Martel's final hour had come. The rope was finally cleared, and the 'Father of Caving' emerged to a ripple of applause from the gathered crowds.

Martel's triumph in the very heart of one of Britain's major caving areas gave the sport the impetus it needed in this country. This spurt was strong enough to lead eventually to a breed of cavers capable of tackling the deepest and toughest underground challenges in the world – even France!

By the beginning of this century, cave exploration was established in the main regions of Yorkshire, Derbyshire, Somerset and South Wales. Looking back at these pioneer days, it is impossible not to pay tribute to the sheer hardinesss and determination of our early cave explorers.

Their clothing consisted of old tweed jackets and trousers, a pullover or two, perhaps a pair of long underpants. Helmets were virtually unknown, the only protection for the head being a cloth cap – almost a compulsory fashion accessory then. Boots were worn by those lucky enough to own a pair, otherwise ordinary shoes had to be sacrificed to the cause.

For many years, candles were the only source of light trusted in the tight confines of a cave passage, hand held or stuck in a cap-band. Herbert Balch, a Mendip pioneer, was a devotee of this form of lighting even as late as 1937 when he wrote: '. . . Candles are by common consent the most dependable illuminant, as they cast no treacherous shadows . . .' A more cynical comment on this attitude was provided by a later generation of Mendip cavers with modern miners' lamps:

> 'Balch's dependable illuminant,
> It is a candle bright.
> It casts no treacherous shadows,
> For it gives no bloody light!'

Balch himself was the victim of another hazard facing the early cavers: rope failure. Before the days of nylon and other synthetic fibre ropes, hemp and manilla was the only choice. Like all natural fibres these are treacherously vulnerable to rotting in damp conditions, and the average caving trip guarantees a thorough soaking of all equipment.

Balch was being lowered down the 65-foot pitch into the main chamber of Lamb Leer, on the end of a rope fed from an old miner's winch. The rope had been stored under a leaky roof and snapped under Balch's weight. His life was saved purely by chance, for one of his outstretched hands grasped a thin cotton line hung down the pitch for anyone to hold if they started spinning on the winched descent. Balch's fingers were cut to the bone by the thin line, but his fall was slowed sufficiently for him to survive.

The rope ladders used then were rather easier to climb than the modern version, but they were terribly bulky and heavy. Old fashioned caving with its complete lack of water protection must have been hard enough, but potholing was even worse. There are probably few cavers nowadays who would care to repeat some of those early trips, soaked to the skin, one hand grasping a candle, the other dragging part of the hundreds of feet of ladder and life-line needed to get to the bottom of even a modest pothole.

To you, hardy gentlemen, we raise our helmets.

Chapter 3

Caves and Cavers

Now that we have examined the basic ingredients for the creation of a cave, it is obvious that when you are looking for caves you go to the limestone areas. In the British Isles there are four major caving areas: Yorkshire, Derbyshire, Mendip (Somerset) and South Wales. An important attraction for the caver to his sport is that all of these regions are of outstanding beauty above the surface as well. With a very few exceptions, caves are found in some of the most attractive landscapes Britain has to offer. No doubt there are occasions when a caver, clambering out of an entrance in his soaked clothing to face the long tramp over benighted and befogged moorland to his hut or car, wishes he could tumble instantly into a scalding hot bath. But the experience of exiting to a gentle hued summer evening in that same country – or even the first probings of dawn after an especially long trip – is more than adequate compensation.

Yorkshire, the county which acted almost as the incubator for British caving, still holds a magical appeal for those in the game. The home counties' clubs in particular regard it with a degree of reverence, and whilst South Wales or Mendip trips are treated on a matter-of-fact basis, the Yorkshire meets are something special.

The potholes and caves of Yorkshire have a well-deserved reputation for toughness, and in turn the cavers of the North were long regarded as the hard men of the sport. Ten years ago, cavers in the southern clubs nurtured private suspicions that their Yorkshire counterparts must be a steely breed, inured to cold and wet after years of plumbing the depths of such cold giants as Penyghent Pot – one of the country's deepest 'pure' potholes at 527 feet – and Mossdale Caverns with its Kneewrecker and Marathon passages. With the opening up of a comprehensive motorway network, and the coercion of more car owners into their membership lists, clubs all over the country have found that they can make regular weekend trips to almost any of the caving regions. In turn they have found that they can confidently tackle any of the major systems without having to tag along behind a local club team. The northern cavers are not genetic freaks of underground supermen after all.

In greatest contrast to Yorkshire's caving region stand the Mendips. Except for the enormous puzzling gash of the Cheddar Gorge, here the land is not one of looming mountains and great grinning limestone outcrops but gently folded hills, like kneaded pastry covered in bright green velvet. Even the names are gentler, more poetical – no becks, ghylls or chases, but such delightful finds as Velvet Bottom and Priddy Green. Although much smaller than the northern Pennines, the Mendips still command a very strong local allegiance, so strong in fact that Mendip cavers seem more content to restrict their caving to their own area than any other 'clan'.

Although no British caving region is in the happy position of offering dozens of open and yet unexplored caves, it is on Mendip that the search for new caves – as opposed to new parts of known caves – has reached the highest pitch. It is a long-standing joke that if a Mendip club cannot find a new cave it simply digs one, but due acknowledgement must be given to the incredible persistence of the diggers there. If there is enough evidence of a worthwhile cave system under their feet, a Mendip club or consortium of dedicated individuals will spend months or even years sinking a shaft or tunnel to reach it. It is true that cavers elsewhere will eagerly blast and shovel their way down to possible finds, but rarely with the long-term determination of their Somerset cousins.

Besides their dogged persistence, the Mendip cavers are probably as equally famed for their above ground social activities. From the smoke and cider-fume filled cavers' rooms in such haunts as the Hunters Lodge, hermetically sealed off from the rest of the pub, have come some of the best items in our caving repertoire.

Somewhere between Yorkshire and Mendip, in character rather than geographically, lies Derbyshire. Here at the more southerly part of the Pennine chain, the Peak District offers quite a rich selection of both caves and potholes and a more equal balance of the two types than elsewhere. Many of the best are within a few miles of Castleton, where the countryside enjoys the protection of one of the best-run National Parks.

To the non-caver, Derbyshire is often regarded as the only main caving region in the country. There are two reasons for this: the Neil Moss tragedy in Peak Cavern, with its prolonged and unsuccessful rescue attempt reported on every front page for a number of days, and the number of show caves.

Whilst touching on the subject, it should be mentioned that show caves often present a problem for the caver trying to explain his sport to someone who has been down a commercial cave. Whilst the latter can in the best cases convey something of the remarkable beauty of underground formations, it is a far cry from the true wild cave. Where the caver has to crawl, scramble, squeeze and swim to make progress, the tourist merely walks along concrete paths and over convenient bridges, perhaps occasionally stooping where the excavators deemed it safer – or more authentic – not to blast a higher connecting tunnel.

But however much cavers might smile at the often hilarious comments made by tourists during and after a tour of a commercial cave with its paths and strings of lights – 'But don't you get *any* light down here when the sun's out?' – they do serve their purpose. Few who have seen the best formations at Gough's Cave, White Scar Cave, Treak Cliff Cavern, Dan-yr-Ogof and others can doubt the purpose behind the quest for cave decorations in their natural state.

Completing the big four of caving areas is South Wales. This is the land of big river caves, from the part-commercial Dan-yr-Ogof with its tight control over access by both the management and the South Wales Caving Club, to the outstandingly attractive Little Neath River Cave – open to anyone prepared to pay a few pence to the farmer who owns the land, and to brave the tiny entrance in a river bed. Even at normal water height the river laps into this entrance, and it takes but a little rain upstream to make the first hundred feet of constricted and twisting entrance passage impassable.

Potholes are rarities in the Black Mountains, but South Wales does boast the longest caves in the country: Ogof Ffynnon first and third Ddu, with over 24 miles of passage, and Ogof Agen Allwedd, nearly 16 miles.

Ireland boasts some very fine and extensive cave systems, and is the objective of an increasing number of summer expeditions by clubs from England. The Irish themselves have caught the imported enthusiasm for underground exploration, and over the last few years have begun serious work in their own right. Both Ulster and the Republic have considerable untapped cave potential, the only snag to extensive exploration there being the relatively sparse cave rescue facilities in the event of an accident. This is generally overcome by English parties taking along their own rescue equipment so that in the event of a mishap they would not have to sit back and wait for rescue teams to be flown over from England.

Smaller caving regions exist in Devon, Scotland, North Wales and the Furness coast. Local cavers make up for the lack of size, though, by sheer enthusiasm in exploration, digging, water tracing, surveying and the like.

These, then, are the places where cavers gather to follow their unique sport, beneath the hills and mountains, moors and fells, by night and by day, under blazing August sunshine or six feet of January snow. And this is another pointer to what sets caving apart from other sports, for unlike conventional activities which are limited to one sort of weather or another unless some form of weather protection can be provided, caving knows no seasons.

The temperature within any particular cave varies only fractionally from winter to summer, apart from those stretches of passage leading immediately to the surface. The study of the mini-climate within a cave is a particularly complex affair, but in general terms the enormous mass of surrounding rock acts as a buffer against any radical change in temperature. It loses heat as the air cools, and absorbs it again if there is a rise in air temperature. This heat-brick effect ensures a pretty cool but constant temperature in most of our caves.

Many non-cavers believe that cave air must be rather foul, bottled-up for thousands of years. In fact it is positively sweet, and most passages and chambers are air-conditioned on a scale which would make envious the designers of modern four-star hotels.

This movement of cave air has two main causes: the rising of the relatively warm cave air through colder outside air, and the sinking of relatively cold outside air, or vice versa according to the season; other more local but often violent disturbances are caused by waterfalls, from a drop of a few feet in a streamway to a plummeting deluge down a shaft hundreds of feet deep. In large passages and chambers the draught is usually undetectable, but where the caver has to crawl through a low section he is sometimes confronted with a miniature gale quite strong enough to blow out an acetylene lamp flame.

Usually just an added cold hazard, particularly when your clothing is wet through, the draught can come as a positive godsend to those patiently digging through a boulder or mud obstruction – for a good healthy draught can mean plenty of cave system waiting to be discovered.

When the beginner crawls out of his first big cave into a warm summer day, something else convinces him of the purity of cave air . . . the stink of the outside. As he tackles the last few hundred feet to the entrance, he will notice that the cold almost clinically clean cave air is being replaced by a warm, fuggy, often oppressive atmosphere laden with pollen, dust and other constituents of what we would regard as 'good clean country air'.

What that same beginner certainly will not miss is the very high humidity of an active cave, an atmosphere so damp in fact that clothing, once wet, simply will not dry unless the caver pushes out quite a lot of body heat by strenuous exercise. Probably the most infuriating caves are those in which one has to slide through a small pool of water near the entrance, even though the rest of the cave may be quite dry. The initial soaking on the way in is gradually dried off, only to be repeated hours later as you crawl to the entrance.

Daren Cilau, a very constricted system initially, plays this sort of trick on cavers in South Wales. So low are the first few feet that the helmet must be removed to allow you to squeeze through, and there is no escaping the two or three inch deep pool right in the middle. It was this very pool which once nearly brought a chilly night's enforced stay on the author and another caver. During our exploration of Daren Cilau the winter night had dropped the temperature sufficiently for the pool to freeze, and when I came to slide those last few feet to the surface I found that there was just not enough room for me between the layer of ice and the roof of the passage – those couple of inches of water made all the difference.

Fortunately the water was not frozen solid, and lying flat out with one arm stretched before me I was able to smash the ice with a small rock. Our wet boiler suits froze crisply within a few minutes on the surface, and we were glad of our early return before the pool had become solid ice. Our rescue would probably have been the first to depend on a blowtorch and a bag of salt.

Let us turn now from the caves to the men and women who explore them.

Cavers come from every social stratum and profession it is possible to imagine. An analysis of my own club reveals a typical mixture of teachers, electricians, students, engineers, civil servants, carpenters, doctors, office secretaries, company secretaries, and so on. This is a sport where the worth of someone is judged solely on his or her enthusiasm, skill and attitude – not on how much they can afford to spend on luxury equipment. The basic equipment is within the reach of everyone, and to be truthful there simply are no luxury items of equipment to be bought; the days of advertisements reading 'Impress your fellow cavers with this supertough ultralight longlife lamp, only £35' are a long way hence.

Akin to caving in many respects, climbing has now entered the high fashion stakes, and the range of mountaineering boots in elegant colours matches those seen at the most *avant-garde* ski establishments. Not so with caving, where the most the ladies can do with their garments is to paint the helmet with some floral arrangement and use a different colour tape for making their wet-suits – hoping, no doubt, that these will distract any onlooker from the baggy boiler-suits and clumsy boots.

Nor has caving undergone the same sort of social change as experienced in the climbing world. There, the sport was once restricted in the main to alpinists who could afford long holidays in Switzerland, but changed character completely between the wars when there was a great outpouring of working men from the cities in the north to the hills. Caving always has been, and presumably always will be, a sport for anyone with the inclination and a few pounds for basic gear.

Whilst there is absolutely nothing to stop someone quite old from tackling caves within their level of competence – Norbert Casteret, the doyen of French cavers, celebrated his 56th birthday whilst being winched up a 1,100 foot entrance shaft in the Pyrénées – the general age bracket of active cavers spreads from the late 'teens to the late thirties. It is a sport which some people try just a couple of times, whilst others pursue it with enthusiasm for dozens of years. The average really active life of most cavers is probably from seven to ten years.

Sailing, skiing and even climbing have obvious attractions for women: they can, if they wish, confine their public appearances to days which are really sunny, and even then not feel committed to do more than a token ten minutes of actual activity. The rest of the day can be spent sunbathing on deck, on the hotel balcony, or at the foot of the rocks, and the evening offers plenty of scope for sporting their latest après-whatever dresses.

It seems strange, then, to find any women at all interested in caving, or actually down a cave. Yet there are precious few clubs without a reasonably large contingent of women, and some of the fairer sex – with smallness and a remarkable resistance to cold on their side – are nothing short of outstanding in modern caving.

It is difficult enough to analyse the motives behind a man's decision to take up the sport, but even more so in the case of a woman. Perhaps many see a mainly male caving club as rich man-hunting ground, but they would find it progressively more and more difficult to avoid committing themselves to actual caving trips. The enthusiasm of the caver to get any available non-caver underground can be quite astonishing, and psychologists would no doubt attribute this – in the case of the caving male and non-caving female – to some basic caveman motivation for the preservation of the species! It must be confessed here and now, though, that the irregular rumours of wild and passionate 'potholing orgies' (for some reason usually centred on Derbyshire) are without foundation. Underground amour, with its attendant difficulties of your being clad in a wetsuit, boiler suit and boots, is far too daunting a prospect.

The presence of a woman tends to have a moderating effect on the often strong language of the male cavers battling away at some underground obstacle, but one realm in which the feminine touch has never been suffered is the club cottage or hut, if the club is fortunate or affluent enough to have established one in a convenient caving area. Once almost *de rigueur* for any self-respecting club, such accommodation has become almost impossible for all but the largest and richest of clubs, so sharply have property prices risen even in remote areas. Now the smaller clubs count themselves lucky if they can persuade some farmer to rent them a lean-to shack or even a chicken house.

It would be unfair to say that the average caving hut reflected the good or bad taste of its occupants; décor must come well down the list when prime items are lamp cell charging racks and storage lockers for ropes and ladders. Construction and renovation is obviously limited to weekends and holidays, so huts generally evolve into unique places, the standard of comfort and cleanliness changing abruptly as you walk from room to room. All depends on the priorities originally decided upon: one cottage boasts a fully equipped kitchen which would delight any housewife, but the bedrooms are a jumble of musty sleeping bags lying on the floor; the next, an asbestos hut dating from the last war, has a dormitory lined with plushly mattressed bunks and chintz curtains at the windows . . . but the queasy avert their gaze as they struggle to cook breakfast amidst the greasy plates of long months past.

So tedious did one northern club find the process of washing-up that they adopted a quite novel procedure. They accumulated vast piles of ex-army tin plates and bowls, and the few that were used for each meal were then tossed out of the window. In theory, one great washing-up session per month should have sufficed, but there were no volunteers for even the first of these. Ten hungry men found themselves sitting round the table gazing at the empty plate racks; outside, in the night, lurked an enormous mound of filthy plates. The idea was dropped.

Segregation of sleeping quarters is something that doesn't often happen in the newer caving clubs, and indeed some might even find the idea objectionable. As a club matures, however, and a fair proportion of its members marry caving girlfriends, it is often considered expedient by the committee – many of whom are the old brigade anyway – to have separate bedrooms for men and women. Some even thoughtfully provide separate quarters for married couples.

Different sports seem to breed different humours, and that of caving has a blunt and often macabre quality about it. Sometimes this is borne out within the caves themselves; many an isolated stalagmite boss sticking up from the floor of some low crawl has been christened the Virgin's Delight, and solemnly recorded as such on the official survey. Propped beside a particular foul sump in one cave is a purloined noticeboard warning 'Bathing is not permitted', and any smiles that this might draw are more over the thought of the desperate struggle the perpetrator must have had in manhandling it down to this deep spot.

It is in the caving huts, though, that real ingenuity in humour is shown. Whitewalls Cottage, the South Wales headquarters of London's Chelsea Speleological Society, lies only ten minutes' walk away from Daren Cilau cave. Rescue practices in this cave have only served to show that a real rescue from this initially very constricted system would be nothing short of a nightmare. It was completely apt, then, when somebody – after a gruelling first taste of the cave – hung a toy pistol on the wall at Whitewalls and labelled it: Daren Cilau Rescue Kit.

Possibly the most ingenious manifestation of cavers' humour, and certainly the most sadistic, was perfected at a Mendip hut some years ago. Uninitiated visitors to the outside Elsan chemical toilet would be intrigued by the small handle sticking out of the wall just below the roll of toilet paper. With time to kill whilst sitting there, and that damnable streak of curiosity in all homo sapiens, the outcome was inevitable – the handle would be turned. Enlightenment was instantaneous. The handle belonged to a small but powerful magneto, mounted discreetly outside the cabin, its two wires taped to the toilet seat. In time, news of this homespun electric chair spread far and wide, resulting in a disappointing success rate. The magneto was moved inside the hut itself, and passing cavers were free to give its handle a quick spin on the off-chance that the toilet was occupied. Sadly, rumours of sterilization brought an end to this quaint version of Russian roulette by proxy.

If one were to seek a reason for this extreme form of humour, it may not be necessary to go further than the supposition that it provides some kind of safety-valve for the sport. A serious caving or potholing trip, particularly in bad water conditions, puts the caver in quite a high state of tension; such an environment is simply not conducive to a completely relaxed state of mind. This is not to say that every caver is a twitching bag of nerves, but rather that he finds his concentration at stretching point for long hours on end, and he may well need some way of unwinding when he surfaces.

Without any doubt the most popular venue for this unwinding is the nearest cavers' pub, an integral part of the whole social side of the sport. I say cavers' pub rather than just pub, for the ideal pub must have certain special qualifications.

The first of these requirements is some small back room into which considerable numbers can be insinuated, and which is acoustically sealed to protect the more sensitive ears of the other customers. The second is an understanding landlord. Here, with the smoke and decibel levels rising to barely tolerable heights, the unwinding takes place. Here is the breeding ground of the new caving songs which at last are usurping the borrowed rugby renderings. Here are the first presentations of the Homeric caving sagas from such muddy bards as Alfie Collins of Mendip. Here is the market place for shop talk and gossip, where furtive hints as to the whereabouts of the latest secret dig have to be screamed above the hubbub. Here is the briefing and debriefing room for which caves are loose, how much ladder you will need in Bloggs Pot, which way to go at the streamway junction, and whether that new blonde is attached or not.

Not surprisingly, there is drinking as well – cider on Mendip or in Devon, the best local brew elsewhere. Besides the more obvious reasons of sociability and relieving stress, there is a third even more pressing reason for the comparatively high consumption amongst cavers: sheer dehydration. Any form of liquid is bulky and awkward to take down a cave, and few people trust the potability of cave streams. Hence the often stronger desire to satisfy one's thirst before one's hunger after a long trip.

Alcohol is taboo in caves, for not only does it reduce the level of concentration but it also leads to more blood being circulated close to the skin – with a consequent higher risk of exposure. Fine in principle, this does not resolve what could be called the Saturday night rescue dilemma. By far the majority of cave rescue callouts come on Saturday evening, often quite late, meaning that – with a few abstemious exceptions – the rescue volunteers are going to be called out from the pubs. Ordinarily this simply results in an added vigour in the early stages of a rescue, but it could conceivably present a more serious problem on the night of, say, the anniversary celebrations of a big local club. On such occasions – fortunately infrequent – the wiser visiting teams will confine their caving trip to a quickly bottomed system of relatively low risk.

The social side of caving is just as important as in any other sport, and in some respects much stronger, for the joint sharing of high risks tends to weld caving teams to a greater extent than, say, tennis club members. It is difficult to practise deception underground, particular as to one's own feelings or fears, so any group of cavers tend to get to know one another's characters, flaws and strong points quite quickly. This is a good thing, for it is really only when some level of mutual appreciation is achieved that a team can work quickly and efficiently without the formality of specific directions from the leader.

This question of leadership is a rather interesting one in caving, for there are three distinct approaches to it. This is in complete contrast to climbing: there, there must be a leader in name and function, for it is he who literally leads the way at the 'sharp end' of the rope. He may have no authority, or inclination, to tell his second man precisely what to do . . . but then the second has little choice but to follow up the same route if he wants to remain on the end of the rope.

The leadership of a party of beginners on a caving trip carries with it an explicit authority. A newcomer might gladly accept the instruction to keep to the left of a streamway because the water is too deep on the other side, but he has also got to obey instructions whch may not appear to have any logic behind them. For instance, a leader may spot that a particular beginner is getting too tired rather too quickly, and call an immediate halt to the trip. He would be failing in his responsibility if he did not. But at the same time, he is unlikely to embarrass the person in question by stating why he is calling the trip off. This is a fairly common situation, and one which must be handled with a degree of diplomacy. This form of cut and dried leadership is also favoured by the services, for obvious reasons, on their junior training programmes.

The situation is far less clear cut with a party of fairly experienced cavers. In this case it is implicit leadership, usually by the organizer of the trip who probably knows that particular cave better than the others. If things go smoothly on the descent and

3. Easter Grotto, Easegill Caverns, in the northern Pennines, with its exquisite inverted forest of straw stalactites. Such formations are extremely fragile

A frozen cascade created by a coating of flowstone over boulders in Ogof Rhyd Sych, South Wales

4. The justifiably famous curtain formations in St Cuthbert's Swallet, Mendip. Like bone china, these fine draperies are light-translucent

ascent, he will rarely have to give any actual instructions, just advice on the best routes and belay points for ladders. But again, if the leader is considerably more experienced than the others, they must be prepared to accept without question any orders he may feel it necessary to give.

Smoothest running of all is the team of highly experienced cavers. There is usually no one leader, either elected or implied, but an almost casual meshing together of the various talents. Those quickest at handling tackle will automatically rig the ladder pitches and handlines; the strongest swimmers will go first and last in deep water traversing; those most deft at handling rope will take charge of the coiling up on the way out; the one who knows the system best will indicate the way, but not necessarily from the front; the strongest man will lifeline the big pitches on the way out after a long hard descent.

The extraordinary thing about the latter system is the way it works when things go wrong. If there is an accident to one of the party, or the cave starts to flood, where quite definite and binding decisions have to be made, someone will emerge as the 'situation leader'. There is no fixed formula for discovering who this might be before the event. Once the different factors have been presented to the team – of relative experience, degree of tiredness, physical and/or mental ability, knowledge of the cave and/or type of situation – a self-selected, but completely respected leader will emerge. This spontaneous emergence does, of course, rely on a high degree of experience on the part of the team, and some considerable time spent caving together.

Large-scale expeditions are rather different. There the stay underground might extend to a fortnight in a really big system, and the logistics of food and equipment carrying are far more complex and demanding than an ordinary ten or twelve hour trip down a British cave. Regardless of the experience of the expedition team members, there has to be a definite leader capable of coping with the many administrative headaches as well as planning the attack day by day. It is common in fact to have two leaders, one in charge of the underground side of things whilst the other controls surface activities to ensure an adequate flow of manpower and materials into the cave. In this sort of situation, the overall responsibility for the expedition is carried by the underground leader as a rule, for in caving leading from the front line usually works most effectively.

Although there are a number of independents, most cavers attach a considerable importance to the caving club system, and membership of a good club does have a number of definite advantages.

For the beginner, clubs still provide the most readily available training in the sport. They will usually find that special beginners' trips are laid on fairly frequently in straightforward systems, and there will be a number of experienced cavers on hand to give the necessary instruction. Quite often, the complete beginner will be able to borrow at least a helmet and light for his first couple of trips, thus saving him wasted expenditure if he decides not to continue with caving.

It cannot be pretended that a would-be caver will always find a welcome at any club, or full training facilities. Some clubs are very conscious of their responsibility, and feel that even if they are bursting at their seams in terms of numbers of members it is not fair to reject a serious application from a newcomer. They are only too aware of the consequent risk of the untrained soul 'giving it a bash' on his own or with a friend. Other clubs, though, have a strong clannish tendency, and even if newcomers are admitted into the official lists they may find it difficult to break into one of the various cliques and get invited on suitable trips.

But if he gets a cool reception at one club, the determined beginner does have a large number of others to try in most parts of the country. The early 'sixties saw the beginning of a tremendous upsurge in the number of British caving clubs, and the most recent figures provided by the *Descent Handbook for Cavers* shows a level of well over 400. The number of cavers in the country is estimated at around 16,000 – a far cry from the popular image of a few dozen scrubby characters roaming the moors looking for holes to crawl into.

Shared transport and regular newsletters are other obvious advantages in joining a club, but perhaps the most important of all is the availability of tackle. A dozen keen members can, in a few weeks of concentrated work at the weekends and in the evenings, make enough ladder to keep the club supplied for a couple of years.

An annual subscription of a pound or so adds up to a big enough kitty to buy the rope which is needed as well; few individuals could afford the amount of ladder and rope needed to tackle the largest of our potholes.

The first moves in offering a standard course of training for cavers were made by the National Scout Caving Activity Centre at the Whernside Manor near Dent, Yorkshire. There one can take courses in cave leadership, cave science or basic caving and potholing. This is a reliable form of training, handling hundreds of newcomers to the sport each year, and quite satisfactory as long as the output of trained cavers can be absorbed into the conventional club network.

Far less satisfactory is the 'training' provided by some schools, youth clubs, scouting groups and the like. Time and time again I, like any other caver, have come across groups of young people struggling through a cave with torches held in their hands, or perhaps only one handtorch to each group of three or four. Perhaps there are one or two sloppily fitted helmets, but often only berets or scarves, and the youngsters fight to keep their balance on awkward ground in utterly inadequate everyday shoes. Instead of the recommended ratio of one instructor to every two or three beginners, there may be only one teacher or youth leader trying to keep control over more than a dozen excited youngsters.

Take one of these 'instructors' to one side for a quiet word of warning and he will usually smile gratefully and reply, 'Oh, that's all right . . . we're not going in very far, anyway. But thanks all the same'. Cavers learn a hard lesson: danger starts the instant you set foot in a cave, however easy, and it is not always obvious. There is no evidence that the deeper you go into a cave the more risk there is, simply that rescue becomes more difficult. A crippling or even fatal accident is just as real ten feet from the surface as it is a mile underground, it is just easier to get what remains of the victim to the daylight.

There seems to be no solution to this problem of well-intentioned amateurs taking youngsters into caves. As long as the current trend of 'character training by adventure' continues to hold sway, open caves will remain a prime target. Faced with very much the same kind of problem in the mountains, and following two particularly widely publicized winter tragedies involving young

children, the British Mountaineering Council recommended that
all leaders of such parties of youngsters should hold the Mountain
Leadership Certificate. A long-established body, the BMC spoke,
and local educational authorities listened. But the caving equivalent, the National Caving Association, has a comparatively short
history and rather less power to wield.

The events which led to the setting-up of the NCA are long
and involved, for the intrusion of politics into caving – not a recent
diversion – has always raised a number of hackles. With the
formation in the mid-'sixties of the big two regional bodies, the
Council of Northern Caving Clubs and the Council of Southern
Caving Clubs (the latter with initially confused motives, some
claiming that it would staunch the 'threat' from the North, others
that it would unite with the CNCC to form a national front), the
caving press resounded with bitter correspondence over the
possible outcome. Without any doubt, access was the main concern
of those who questioned the regional councils: would they, whilst
under the guise of easing the access situation by negotiating with
farmers and landowners to let them run access arrangements,
make the situation even more intolerable with masses of red tape?
Today some would claim that they have, particularly in the
North.

The access situation is a very involved and complex one and we
are a long way in Britain from any kind of national policy, whether
one of negotiating at a regional level – which is what generally
takes place – or one of national negotiation by the NCA. With
conservation, access is the caver's biggest problem today. Very
many cave entrances lie in private property or state controlled
land – and it is the entrance only which matters to the caver, for
obvious reasons – which means that access to a cave all too often
depends on the patience and goodwill of the landowner. Where that
patience or goodwill runs out, perhaps because of damage to dry
stone walling – whether caused by cavers or not – it is all too easy
to seal a cave entrance with a small charge of gelignite or a few
bags of cement.

Generally speaking, most access problems are best dealt with at
the local level. Regular cavers in the particular area may well be on
first-name terms with the farmer, and they will have a real understanding of his problems and grievances. Where they may well be

able to work out a solution over a pint of beer, the reaction of the same farmer to an official approach by some stranger from a national organization may predictably be one of reaching for the shotgun. Thus, despite a number of particularly vexing problems, the policy so far of non-intervention by the NCA is a sound one. But there are two quite clearly divided opinions as to the future: some cavers want the NCA to be modelled far more on the lines of the National Speleological Society in America where there are few individual clubs but many 'Grottoes' of the NSS, and con-sequently a strong national spirit about the sport. Other British cavers would prefer the NCA to remain at least as innocuous as it is, without any official 'teeth' over access, training, arbitration or certification of caving competence.

One area in which the National Caving Association *must* concentrate its powers is public relations.

Caving is very much the hidden sport, with its followers tucked away beneath the earth far out of sight of any spectators. This undoubtedly lends a great deal of attraction to the sport, parti-cularly for those who seek comparative solitude with just a few chosen companions. But at the same time it has resulted in the worst possible public reputation any sport could have.

If I may be forgiven for drawing a comparison with climbing yet again – such comparisons are useful, though, for the two acti-vities are so similar in many respects – it is interesting to note the distinct improvement in the public reputation of rock climbers and mountaineers over recent years. Early mountaineers of the last century had to disguise their Alpine clamberings by pretending to be searching for mountain butterflies or plants; this went some way to mollifying the critics of the pastime, but a strong stigma remained well into the 'sixties. Impressed by the earlier conquest of Everest by a British expedition, the public were then treated to a new sort of television spectacular, nerve-tingling rock climbs broadcast live into millions of homes.

At first, viewers of these small screen epics waited minute by minute for the first climber to plunge to his doom. When, after a number of such broadcasts, there had been no gory deaths, the audience became more interested in the actual climbing techniques being shown. Small talk was punctuated by such comments as, 'Joe Brown did some fine stuff yesterday. Did you see him climbing

out of that overhang? Very difficult, that'. Climbing has become
a standard television item, either live or in filmed documentaries.
And, like anything exposed to the public gaze for long enough,
it has been accepted.

Far from being accepted, caving continues to be regarded as the
leper of sports. True, there have been a very few television documentaries based on it, but the difficulties of filming underground
are such that it is next to impossible to convey the 'feel' of a caving
trip. As to live broadcasting, we are a long way from this. A single
climb on a rock face can, with careful siting, be covered in its
entirety by a handful of television cameras; the twisting confines
of a cave offer no such ease of coverage.

With so many problems of film and television coverage, one
might have thought that caving would slink along unnoticed, and
be allowed to continue behind its cloak of rock. As the reader will
be well aware, though, this is not the case.

If a news editor of a national newspaper were asked to define
the perfect story, he may well list his requirements as these: plenty
of human interest; readership interest maintained over a couple
of days; high dramatic content, preferably with a human life at
stake; relative ease of reporting; good photographic subjects; no
hefty payments for publication; at least one aspect for the reader
to criticize bitterly, so becoming emotionally involved. Is it any
wonder that caving has such a reeking reputation when one realizes
that the average cave rescue contains all or nearly all of these
ingredients?

Splash Press coverage of cave rescues is something the caver
has learnt to live with – he has had no choice in the matter,
anyway. But it has generated within him a deep contempt for the
Press, a contempt which, it will be seen, has exacerbated the
situation in some cases.

The job of a newspaper reporter is to report, and he is told
exactly which stories to report on. Whether one likes it or not, a
cave rescue is a good story. A life is at stake, ordinary mortals are
taking risks to save it, the drama may well be strung out over a
few days. It is a fairly easy type of story to report on, too, for apart
from the problem of getting to a telephone to send his copy, all
the reporter has to do is wait by the cave entrance. All the juicy
news he could desire is coming out of that one hole in the ground

. . . if he can get it, for remember the average caver's feelings towards the Press! If the bona fide rescuers will not comment on how the rescue is progressing, the reporter has no choice but to collar anyone who is willing to talk. However, the muddy-faced youngster who tells him of the 'living hell' underground may have done no more in the rescue than empty the WVS tea urns.

To prevent distortion of facts, it is imperative for each rescue organization to accept that the Press will be there looking for information, and to appoint someone as a Press Officer. And that someone must know just what the Press requirements are; it is of little use making a general statement every couple of hours when each reporter has to try and get some slightly exclusive angle. The Press Officer must realize this, and see to it that each reporter is steered to different individuals who are competent to comment on the rescue.

Happily, we seem to be over the days when newspaper stories were based largely on the miners, firemen and policemen who were risking their lives underground to get the victim to safety. Cavers, like climbers, look after their own when something goes wrong, and the Cave Rescue Organizations of the different regions are composed solely of volunteer cavers who are quite willing not only to rescue trapped or injured cavers, but also to train regularly to this end. The police are responsible for the initiation and general running of a rescue by their own insistence, but it is cavers and cavers alone who should be underground. I say should, for in some rescues in the past miners and policemen have gone underground, but not at the request of cavers. Indeed, considerable doubts have been expressed by cavers when the police have on occasions allowed miners to assist. The caving world can never forget the classic quote from one South Wales miner, given pride of place in many newspaper reports, to the effect that 'It's terrible. There wasn't a single pit prop in the whole bloody cave, man.'

Assistance in a cave rescue by completely well-meaning but equally inexperienced people is a rarity nowadays. We thank you sincerely – but let us keep it like that.

How valid is public criticism of caving, based as it is on front page newspaper treatment of rescues? I think the answer is given by another question: if all football matches, boxing matches and rugby matches were conducted inside huge concrete domes, completely away from any spectators and with no television or film coverage, what would the public and Press reaction be to the inevitable stretcher case injuries dragged out of the dome in every two or three matches? Without an understanding of what these sports offer despite the risk of injuries, would not that reaction be similar to that evoked by the hidden sport of caving?

Chapter 4

The Hardware and the Software

It is all very well to praise caving for its lessons in self-sufficiency and freedom from the nightmarish technical trappings of modern civilization, but one does need a certain amount of equipment in exploring the underground environment – and that equipment has to be of a very high standard indeed. Whilst most sports seem to have manufacturers going through every advertising contortion in the book in an attempt to persuade the market to buy their products, caving has never enjoyed a choice of products – indeed, there are precious few items of clothing or equipment specifically designed for cavers.

Perhaps it is because of the tainted reputation caving seems to have acquired along the line, or perhaps because manufacturers are simply not aware of the size of the market. Whatever the reason, cavers have had to develop quite remarkable powers of improvisation and scavenging in order to satisfy their rather unique demands. When the need arises for some new item of equipment, he puts such materials as candle wax, ammunition boxes, polystyrene ceiling tiles, epoxy-resin adhesives and plastic sheeting to uses never dreamed of.

As he stands, Man, the naked ape, seems particularly ill-suited physically to explore caves. He cannot see in the dark, his skull is vulnerable to quite small clouts, his bone structure prohibits a free-fall descent of a shaft, and he gets cold ridiculously soon. These are the basic problems.

The solutions cavers have come up with are, like all good solutions, basically simple.

Above all other appurtenances, the head is that to which most people seem particularly attached, so it is to that awkwardly placed projection the caver first turns his attention. Early cavers seemed content with their cloth caps and perhaps a little bit of extra padding underneath, but however rakish, this form of headwear offers little protection in a serious fall against a wall or projecting boulder, or against the rare falling rock. Quite quickly it was realized that the miner's helmet had much to offer: better protection against impact, and somewhere to attach one's light to leave the hands free.

The first helmets used by cavers were made of a compressed material akin to papiermâché, and whilst these were fine on the surface or in a dry cave system, a long wet trip would reduce them to the consistency of stiff dough – still strong enough to support the lamp, but giving no more head protection than a couple of slices of wet bread. That cunning borne of the caverns resulted in a liberal application of gloss paint to the helmets of those cavers who really knew their stuff, and floral or geometric flourishes of the brush brought the first hint of the psychedelic to the subterranean.

This branch of cave art died, however, with the introduction of glass fibre and plastic helmets. Much stronger than the old helmet, these are also quite impervious to water, so much so that on occasions they double as buckets in the baling of sumps. Cavers become remarkably fond of their helmets, bestowing on them the same kind of rough affection a master gives his dog. To the smoker, they serve yet another function of protection, a sort of last bastion of defence for his precious packet of cigarettes. When two or more smokers stop for a rest underground, the first one to remove his helmet is performing a mime version of the spoken 'Here, have one of mine'. The skill of the smoking caver in negotiating deep water is without equal; he knows that the polythene bag around the packet stowed in the very crown of his helmet will remain waterproof for only so long, and his contortions to keep that part dry in crossing a deep lake or crawling through a threequarter flooded passage have the deftness and grace of a ballet dancer. Such is the varied rôle of the caver's best friend.

Next in the order of protection priorities come the feet – at least for the British caver, who shares his countrymen's concern for those corn, blister, chilblain and cold-bedevilled companions.

Shoes are out of the question, so boots it is. The quest for the perfect caving boot is as dedicated as that for the Holy Grail, and every caver must pass through years of searching, through ex-army

stores and manufacturers' catalogues, before coming to the conclusion that it does not exist. This is no deterrent, though, for after
a year of gloomy disenchantment he will return to the quest, convinced that *someone* must have designed and made one in the
interim.

This is the difficulty: even if the water in a cave is shallow, your
boots will quickly be soaked through and through. In the absence
of readily available synthetic material, leather boots are the only
ones the caver can make his choice from, and saturated leather is
quickly worn down by constant kicking and scuffing against rock.
Even more serious is that stitching rots or tears through the leather
in quite a short time, leaving gaping holes in various parts of the
boot, and securing screws coming adrift result in the sole dropping
away from the boot like the lower lip of some storm-weary whale.
The question of whether to repair the holes in one's boots is a
hotly debated one in caving circles, some claiming that whilst it
does allow the water to flood in, a properly sited hole does allow
the water to flow out again when the foot is lifted from the water.
Such are the finer points of caving expertise.

These problems of thoroughly saturated boots are capped only
by the problems of the same boots when completely dry. The ap-
prentice to caving is often puzzled by the pre-trip ritual of boot
soaking. Warned by the old hands to keep as dry for as long as possible within the cave, he sees these same apparently sane men standing in a pool or stream outside the hut for minutes at a time, hands
in pockets as they discuss the weather or some such topic. When he
comes to prise his own boots on, and particularly when he grapples
with the lacing, he realizes the ritual has practical rather than religious foundations. Unless nurtured lovingly with molten dubbin
or other leather treatment, caving boots after a wet trip will dry
out to cast-iron stiffness, freed only by another soaking. Cavers
who have omitted the soaking ritual from their preparations have
been reduced to near tears after a long approach march to the cave
entrance.

That long expanse of body between the boots and the helmet is
the region which has witnessed the most radical changes in caving
equipment.

Dry caves require straightforward warm woollen clothing of
pullovers, long-johns and the like, but the experienced caver will

more often be tackling systems where a dry stretch is a luxury. If he keeps on the move, woollen clothing will be adequate for an averagely wet cave, but in anything more severe he will face a very real risk if the going gets slow or stops altogether for some reason. Then, the body heat which has kept him going will rapidly diminish as the water in his clothing evaporates. More than anything else, body temperature defines the endurance barrier in this sport.

Early attempts at beating this barrier were based on the principle of keeping one's woollen clothing dry. The late 1950's witnessed the brief 'writhing' era of the drysuit. Those cavers who could afford them would wriggle and worm their way into ex-naval dry diving suits, entered via a sort of misplaced umbilical tube in the chest of the garment. The tube was then twisted into a waterproof knot, and rubber gaskets at the neck and wrists kept the person encased inside in a fair degree of dryness – except for the pints of perspiration which poured out at the least out-of-water exertion.

Then came the poor man's drysuit, the ex-airforce anti-exposure 'goon suit'. These Frankensteinian creations were the undoing of many a good motorist who happened to be passing the roadside changing room of a caving team in those days. In gay primrose yellow, the goon suit had been designed to fit any size of airman who might find himself in the drink bobbing along in a dinghy. The result resembled nothing more than the spare skin of an eight-foot-high Quasimodo.

Unlike the forced entry technique of the naval drysuit, cladding oneself in a goon suit was comparatively simple: one merely clambered in through the huge neck opening and pulled the drawcord tight. The grotesque balloons at the bottom of each leg covered any size of foot, with or without a plaster cast, whilst the wrists were closed by an ingenious and completely ineffective strap and press-button system. For the male caver, though, the goon suit required the addition of yet another exit tube, situated at a critical distance below the navel. The variety of unpatented methods of watersealing this additional piece of tailoring was truly wonderful.

After years of cold caving in sodden woollens, the caver could now begin to tackle the very wettest systems without the constant nagging fear of cold exposure. The goon suit gave him the courage to plunge into any number of icy underground lakes and rivers, and most of the perspiration vapour was vented through the neck opening. There were, of course, snags. The design of the goon suit was such that anyone over five feet in height had considerable difficulty in walking with anything more ambitious than a mincing gait. The thin fabric was prone to snagging and tearing, even under the outer protection of a boiler suit. But worst of all was the consequence of forgetfulness. I recall only too clearly the anguish of one friend as he plunged into one of the coldest lakes either of us had ever encountered in a cave – having stopped earlier in a side passage on a mission of relief, the now stricken man was learning the bitter consequences of leaving that vital additional tube unsealed!

By far the biggest breakthrough in cold protection for the caver was the introduction of the neoprene wetsuit. Again no originality can be claimed, for skin-divers had been using these for some time before the caving clan realized their worth in their own sport, in the early sixties.

The principle behind the wetsuit is simplicity itself, for rather than trying to keep the water away from the body it does just the opposite – so minor tears and rips are of little consequence. Fashioned from flexible neoprene foam rubber, the wetsuit soaks up the water and holds it, in its thousands of tiny cells, against the skin of the wearer. Entry of the water through the front zip and other joints is fairly slow, thus eliminating any sudden cold shock, but the body heat soon heats it up to a mildly warm and quite comfortable temperature.

As with all innovations, cavers had to put the wetsuit through a breaking-in period before its full potential was realized in underground work, and in the case of my own first suit the breaking-in was on the part of my own body rather than the garment. I reasoned that if a skin fit ensured maximum insulation then the wetsuit, being made of flexible neoprene rubber, should be made slightly undersize. This was a mistaken assumption which the rest of my club members took to heart, for it was their services which I had to request to help me dress before any caving trip. Despite liberal application of French chalk, which acts as a sort of dry lubricant, it was sheer hell pulling the tenaciously clinging rubber cells over my arms and legs. When that was done, I would find the hut changing room empty of my closest friends – only some unknowing visitor left to help me with the last stage. That last stage, doing up the jacket zip, took at the very least half an hour. Time and time again the fatigued zip teeth would burst undone half-way down like the skin of an over-ripe banana. Once I was encased in the black shroud, there I stayed until the very end of the trip, each breath a conscious inhalation against the restraining tension of the jacket.

Whilst minor tears are of no real concern to a wetsuited caver, it is important to make the suit with doubly glued and taped seams.

Tich Morris, one of the best and toughest cavers ever to emerge in Britain and now exploiting his speleological talents in the big systems of Canada, could never quite get those seams right. Reinforced taping seemed rather fiddly to him, too. The result was the Morris Box, a battered margarine carton which followed him on every trip up to Yorkshire or South Wales. Bouncing in the back of the club's hired minibus on the journey up to the caving area, Tich Morris would pull shreds of neoprene from his box and hurriedly glue them together into some semblance of the human form. As this operation was done in the middle of the night, the resulting wetsuit gave rise to many rumours of a visiting London caver who was deformed in several startling respects. After each cave, Tich would step from the exit and unconcernedly peel the once more separated shards of rubber from his body, dropping them one at a time into The Box.

Today, cavers enjoy the luxury of nylon-lined wetsuits, vastly stronger than the non-lined neoprene and much easier to put on. So clad, he has pushed back the endurance barrier in wet caves to the point where water is of secondary importance to tightness of particular sections or depth of pitch in a pothole.

'Look after t'lamp, lad,' the wise northern potholer advised me in my early scramblings underground; a sentiment expressed, a shade more poetically, by Ben Jonson with: 'Suns, that set, may rise again; But if once we lose this light, 'Tis with us perpetual night.' Both add up to the same message: men are creatures of the day, and a caver without his light is utterly helpless in the absence of companions.

Which is one sound reason for never caving alone, of course, for it is not uncommon for one man in a team to have to share his light with another after some lamp malfunction. The obvious solution of carrying a spare lamp would seem to be too obvious for some, and they prefer to risk a lamp failure. Sharing a light presents no problems over much cave terrain, of course, but this is obviously impossible on a ladder climb where one at a time is the rule. Woe betide the unilluminated one then if, half-way up a 200-foot pitch, he finds his lifeline knitted through the ladder rungs and has to untie to sort the mess out.

After the era of candles clenched in teeth or shoved into hatbands, the acetylene or carbide lamp held sway for many years, only fairly recently being usurped by miners' rechargeable electric lamps. The swing to electric lighting has been most pronounced in British caving, for 90 per cent of American cavers and many on the Continent still use the faithful old 'stinkie'.

The greatest advantage of the carbide lamp is its duration; given an adequate supply of water and calcium carbide, it can be recharged time and time again underground, each fill lasting for three or four hours. The lamp has two small brass chambers (plastic in the latest USA type) screwed together, the top one containing water, the bottom one calcium carbide. A tap allows the water to drip slowly on to the carbide, and acetylene gas is given off through a small jet in the centre of a polished reflector. This results in a cheerful flame, useful for hand-warming or cigarette lighting as well as casting a pool of soft light, but this exposed flame is the lamp's weak point in waterfall scrambles, sumps or on wet ladder pitches.

The use of carbide lamps certainly adds a new dimension of hazard to caving, as at least one caver found to his cost in Agen Allwedd, South Wales. He was given charge of a team's five pound tin of carbide, and grew a little concerned at the increasing smell of unburned acetylene which had pursued him for the past couple of hours. Opening his haversack, he took out the tin and prised off the lid to check for leaks. Obviously he had never heard the old joke about looking for a gas leak with a match, or he might have thought twice about peering into the tin with a carbide lamp burning merrily away on the front of his helmet. The resultant explosion did not damage his eyes, by some miracle, but a complete lack of eyebrows, eyelashes and sideburns served to remind him of his folly for some weeks to come.

Apart from the occasional fried finger in some particularly restricted fissure, my own carbide lamp damage has been confined to anatomies other than my own. To this day petite Patsy has been unable to forgive me for the atrocity I perpetrated in the non-tourist extension to Speedwell Cavern. Her helmet removed as she tried to get her own carbide lamp working properly, she smiled sweetly at my offer of assistance, her face framed by a slightly muddy but undoubtedly expensive coiffure. An expert clout or two re-sulted in a brightly burning lamp, but as I raised my eyes I saw Patsy's fringe burning equally brightly from my own lamp flame. She thanked me sweetly, unaware of the pyrotechnics a fraction of an inch above her scalp, then fell back speechless as I smacked her sharply across the forehead without time for an explanation.

The days of the carbide lamp illuminating the underground scene with a cheerful localized glow are numbered, at least in Britain. More and more cavers are turning to electric lighting as ever wetter caves are explored and the conservation-minded stress the possible dangers to cave life if spent carbide is dumped under-ground. There is the additional drawback of unsightliness when the more thoughtless cavers do not bother to bury their spent car-bide, or better still take it out of the cave with them, but merely dump it where and when they stop to refuel.

The light source which is illuminating the way for today's cavers is the miner's-type lamp, with its rechargeable accumulator. The three basic makes, Nife, Oldhams and Edison, will give light for ten to twelve or even fourteen hours, depending on their owner's con-cern for maintenance and careful charging. It is not uncommon for newcomers to the electric brigade to find themselves finishing a long trip in the near-dark, but they quickly learn the trick of only using the main beam bulb when strictly necessary, the rest of the time relying on the half-power reserve bulb. The real masters of light conservation are never seen with their lamps on if they stop for more than a minute, and often find the light from other lamps quite adequate for a straightforward passage. The lamp masters know how to groom their light sources to get the best possible per-formance from them in the underground environment: the by-passing of the fuse to avoid any sudden blackout; the gluing of a toothpaste tube cap to the low-profile switches, awkward to turn on with cold and muddy fingers; the substituting of plastic gas-venting caps with leakproof steel ones (electrolyte burns from leaking cells have scarred a few cavers for life).

With their stainless steel or tough plastic cladding, these lamps – weighing about eight pounds – look indestructible, but this is a word to use with caution in caving. One caving companion was particularly proud of his newly refurbished Nife cell, and could not believe it when it failed completely towards the end of one particular trip. We had completed a through-trip of the notorious Hensler's Passage in Yorkshire's Gaping Ghyll system, a trip which involves some thousands of feet of often flat-out crawling over sharply rippled rock. He had not bothered to keep pushing the cell on to his back, and the harsh rock floor has scoured away one corner of the stainless steel case and that of the inner cell casing. Exit electrolyte – and light.

Spare or emergency lighting has been the subject of much inventiveness, and can vary from a single candle (edible, if you want to be really safe!) to the ingenious waterproof torch designed by the Speleo Rhal Caving Club of Southampton. Little bigger than a matchbox, this is powered by two small alkaline batteries which can power the lens-ended bulb for several hours, and the whole unit is first wax dipped and then sealed in epoxy resin. The foolproof switch is a reed-switch operated when a tiny magnet is pushed into an aperture in the case.

Carbide lamp users often have a dry-battery lamp as both emergency lighting and main-source lighting for extremely wet conditions. Normal dry-batteries do not live up to their name for very long in a cave, and can be 'custom ruggedized' by dipping a bell-type battery into candle wax and wrapping the whole unit in a layer of insulating tape. I found that such a unit will have a vastly extended shelf-life, and not disintegrate even after total submersion. The same, alas, cannot be said for the caver.

When one considers that a single trip can wear through a steel lamp case, it is not surprising that the less well-padded caver, in terms of fatty tissue, will go to extraordinary lengths to protect his elbows and knees from the constant jarring and scraping of a long crawl. An extra pad of neoprene stuck on to the wetsuit will usually suffice, but for ultra-long crawls like Hensler's or that in Penyghent Pot, kneepads are a blessing. None is specially fabricated as yet for cavers, so they make do with the gardening type, or strike up a friendship with a road mender – or two, for they often kneel on the same knee all day long, and use only one pad.

There are two schools of thought on wearing gloves in caves: some use them all the time, while others prefer to let the skin harden in its own time. One caver struck me as particularly fussy about his hand care during trips, then I remembered his daylight vocation as a surgeon

This, then, is the basic equipment of the caver (costing about £45 at 1975 prices, including a wetsuit and accumulator lamp), all attached to the body or clothing apart from the small 'spares' bag, leaving him otherwise free to crawl, climb or walk relatively unencumbered. What makes potholing so much more difficult, particularly for the small party of six or less, is the amount of tackle which has to be carried to bottom a deep pot with many long pitches. Modern tackle is undoubtedly lighter and less bulky than in the early days of the sport, but this is small consolation when one is snaking along the thousand foot crawl on the way out of super-severe Penyghent Pot with a tackle bag tied to one ankle and dragging along behind for all the world like some convict's ball and chain.

Old-fashioned rope ladder has almost completely disappeared from caving, except for the odd surface shaft tackled by some club with a penchant for nostalgia. The big wooden rungs, although a delight to climb in clumsy caving boots, were far too bulky; the side ropes absorbed water, adding considerably to the weight and presenting the threat of rotting fibres if not dried out properly.

The descent of a shaft today is accompanied by a gentle metallic tinkling, the music of rung against rock. Modern rungs are fabricated from exceptionally strong aluminium alloys, pencil thin and only wide enough for one boot. Rope sides have been replaced by fine steel cables; these pass through holes in the rung ends and are secured by steel pins and epoxy resin adhesives, or one of a number of other methods. Different lengths are linked together to meet the particular depth of each pitch, and the top is belayed by steel cable to the strongest available anchorage. Where a natural belay point is absent modern expansion bolts are used, or pitons knocked into some accommodating crack.

The golden rule of security on ladder pitches is always to use a lifeline, however short the pitch. But, like many golden rules, this

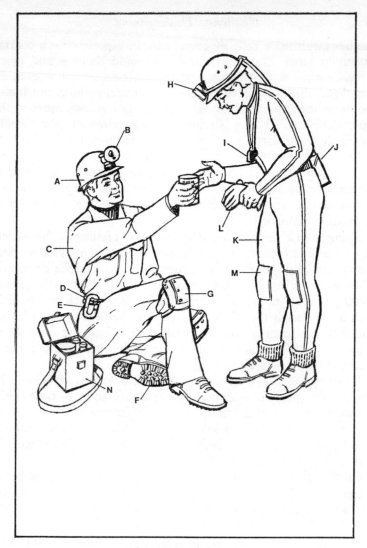

Personal equipment

A – Glass-fibre or plastic helmet
B – Carbide lamp
C – Boiler suit covering woollen pullovers, long johns etc.
D – 20-ft nylon waistlength
E – Karabiner
F – Cleated rubber-soled boots
G – Kneepads

H – 2-bulb switching cap lamp
I – Signalling whistle
J – Rechargeable accumulator
K – Neoprene wet suit
L – Gloves
M – Knee reinforcing
N – Spares bag with spare light, food, drink etc.

one has tarnished a little as cavers gain in experience – a contradiction in terms the safety-conscious would claim – and many teams dispense with lifelining on pitches less than 20 or even 30 feet. True, this does speed up the descent of a pothole, but it also speeds up the descent of anyone who makes a clumsy move at the top of the pitch. Whilst a fall down a one hundred foot shaft would invariably prove fatal, a fall of 30 feet may well result in a very mangled, but still living, caver, and a candidate for immediate cave rescue. There seems to be a feeling with some that ladder risk only starts above the 30 foot level, and as one who has dispensed with a lifeline myself on short pitches when time was short, I wonder at this ability to fool one's self.

Tying on to a lifeline is no insurance against falling off the ladder; only when that lifeline is under the firm control of the lifeliner does the system begin to work. The rope – nylon, terylene or certain other synthetic fibres giving a breaking strain of something over 4,000 pounds – is passed behind the lifeliner's back and held in such a way that if the climber falls he can be held by the lifeliner slamming the arms across the chest, thus locking the rope. Most vulnerable of all in his eyrie at the top of the shaft, the lifeliner is secured to the solid rock by a separate belay.

Unless the team is big enough to be able to leave a lifeliner at the top of each pitch, the double-rope system is used to protect the last man down – and the first one up. This means that the lifeline, rather more than twice the length of the pitch, is threaded through a pulley at the top of the climb and a lifeliner can then protect the climber from the base of the shaft.

Climbing a big pitch on electron ladder, especially where it hangs free of the rock, is a unique experience. There is absolutely no comparison with climbing a rigid ladder, for the caver's steel and alloy version twists and turns as the weight is transferred from rung to rung. If both feet are placed on the ladder from the front only, the electron ladder will push out from the climber at an angle, throwing most of the climber's weight on to his hands and arms. This can result in sudden and violent cramp, and the climber being unable to hang on. The technique evolved over the years by cavers on electron gets the weight back where it belongs, on the feet: hands are wrapped behind the ladder in front of the climber's face, and one or both the legs are also bent behind it, the feet being placed on the rungs from that side. This technique ensures an up-right, safe stance. Obviously, when the ladder is against a rock wall the feet and hands have to hold from the climber's side, but the closeness to the shaft wall prevents any bending of the ladder anyway.

Yet another technique has had to be evolved where the rungs – pencil thin, remember – are hard against the rock and give no real purchase for the feet or hands: on the ascent, the boot actually on the rung holds the ladder clear while the rung above is located; on the descent, the leg holding the climber's weight on a rung is bent underneath him slightly to pull the rung below clear of the rock. Both these methods also keep the ladder clear enough of the rock for the fingers to get behind each rung.

Clubs which do not have potholes within easy reach have to hold their ladder practices in sometimes rather unusual surroundings. When my own club was training for an expedition to Triglav Pot-hole in Yugoslavia, which we knew sported a gigantic shaft of un-determined depth, we persuaded the London Fire Brigade to let us hold regular training meets on the 80-foot-high Wandsworth Fire Station tower. Passers-by gazed in amazement as various scruffy individuals practised a fire-fighting technique which seemed to consist solely of climbing up a flexible ladder and then down again. Such training is invaluable, however, and at the peak of fitness several of us were able to ladder climb that tower up and down eight times each way without a single pause.

But it is underground that ladder climbing counts, when you stop and see the tiny pinpricks of light two or three hundred feet below, when the walls are too far away for your own light to illuminate, when you pray – on the way up – that the lifeline did not drop through the ladder. If it did – and, loose rope being a vindic-tive animal, it will have done if given the chance – you must then perform one of the trickiest ladder manoeuvres imaginable. After securing yourself to a rung with a karabiner (a strong metal clip with a spring-gate on one side), you must then untie the lifeline and pull it out of the ladder before retying it to your waistlength. But beware, let it slip from your grasp and you may be left stranded in mid-ladder, the end of the rope dangling just beyond reach. Getting the rope back to a climber in a situation like this can be a nightmarish activity if the top of the shaft is constricted, or the lifeliner is slightly to one side of the direct line of descent.

Even when firmly secured to the lifeline, the caver half-way down a big shaft can suddenly feel the most vulnerable creature, on or under, the earth. At times like this he freezes to the ladder, his former steady climbing pace of rung by rung by rung completely broken. He stares at the cables holding him in mid-air, cables no thicker than parcel string, and as they are but inches away from his face it is easy to pick out the individual steel strands of each one.

He tries to remember: was it an elephant these two gossamer strands would support, or only half an elephant – or even half an elephant each? My God! How much does half an elephant weigh? And doesn't it rather depend on whether the elephant is passive or not?

The cables shrink under his gaze as he becomes acutely conscious of his body weight, and desperate attempts to 'think light' fail miserably. He feels more elephantine with each long second, and cannot brace himself to break the trance by stepping down to the next rung, for surely the slight bounce will snap the cables in an instant. Cables? No, these twisted strands are mere wires. One simply refers to them as cables because that sounds so much stronger. The near-paralysed caver soon reaches the stage of recalling each step of the club ladder-making he helped in. Did he stir the epoxy resin adhesive for long enough? Did he tap home the steel pin? When heat-cutting the wire ends of this length of ladder did he accidentally heat too much of it and destroy the strength-giving temper of the steel?

This phenomenon of ladder seizure is usually broken when the caver's attention is drawn to the gentle twitching of the lifeline as the patient lifeliner tries to 'feel' the progress of the man on the ladder. This rope at least looks fat and healthy and capable of holding up any number of elephants. A small piece of self-deception, but enough to get him climbing again, his confidence restored in these ultra-lightweight steps to terra firma.

On some pitches, even quite long ones, instructions shouted by the climber can be heard quite clearly by the lifeliner, but others have acoustic properties which destroy the meaning of each shout as effectively as a security telephone scrambler, or falling water drowns them completely.

Communication is obviously vital, for although the lifeliner can usually feel the progress of the climber by the tension or slackness of the rope, there are times when this is not sufficient. If a caver is climbing up a ladder and has to clip on and untangle the rope from the ladder he may well need a few feet of extra rope; if this instruction is not received by the lifeliner, he may well pull in all the lifeline until faced with an embarrassingly vacant end. If the rope passes through a crack in the edge of the shaft, the lifeliner may mistake the tension caused by this for the weight of the climber, when in fact there may be some yards of slack rope – and a fall here could break the rope.

The everyday answer to this communications problem is the humble tin whistle, the raucous sort favoured by football referees. A standard code answers most situations: one blast, stop; two blasts, up; three blasts, down. Additional codes can be made up on the spot underground to meet special contingencies.

In the really big shafts, though, even the shriek of a whistle can be lost. In such cases a telephone is the only answer, but when carried by the ladder climber this has its own problems of the wire tangling with the ladder or lifeline or both.

One of the tensest moments during the Triglav expedition mentioned earlier came when three of us were preparing to exit after three days camping underground at the bottom of the Big Shaft. The first of our three kitbags was being winched up this 365-foot rock-and-ice shaft by two of the relief team pedalling furiously on 'Kuratz', an apparently vulgar Yugoslav name which we had bestowed upon our Heath Robinson – but remarkably efficient – rope winch.

The only communications link between our shaft base camp and the vital ledge from which we lifelined and gear-hauled for the Big Shaft was a slender telephone wire. Without that link the three of us had no way of communicating when we were tied on to the life-line and were ready to climb. Whistle blasts could be heard at the top of the Big Shaft, but not through the ice barrier through which the first 60 feet of ladder dropped down a narrow ice passage.

The telephones were completely self-contained transistorized units, and I was speaking into the one by our base camp tent when it was suddenly whisked from my grasp. The time it took us to realize what had happened was too long, and as we dashed to the foot of the ladder, right in the line of fire of the huge ice lumps frequently dislodged at the top of the shaft when winching was in progress, our hearts sank. The kitbag had snarled with the telephone wire, and the telephone itself – pathetically bleating its call signal – was out of our reach and slowly rising on its 365-foot journey.

Some sixty feet from the top of the shaft, the wire, kitbag, hauling rope and ladder suddenly met in the most appalling tangle, and winching stopped.

Powerless to do anything from the bottom of the shaft, we could only watch the eventual descent of Gus Tanner from the ledge to the ugly knot of tackle, wire and rope. Clipped to a rung, with the temperature at freezing point, he needed every ounce of his strength and patience to sort out the mess. Dangling 300 feet from the ground, this took him over one and a half hours before that vital telephone and wire could be lowered to the three subterranean castaways. Fortunately for us, Gus was a strong and patient caver.

Whilst ordinary telephone equipment needs one or two wires to operate, with the receiving and transmitting sets plugged in to either end, newer induction apparatus is being shown as a versatile addition to underground communication. A transmit/receive set need only be held close to the wire for a link to be set up, and this is particularly useful on cave rescues where there is little time for the rescue team to stop and splice a connection into the wire.

Quite considerable distances of rock have been covered by nonwire induction systems. The speleo-technician Harold Lord achieved a remarkable speech link through 3,000 feet of rock at the Gouffre Berger, but he had to use wire coils of nearly one hundred feet in diameter. This problem of the necessity for huge aerials in either long-wave radio or in induction communications means that we are still a long way from the time when speech communication through the solid rock surrounding caves is a matter of course. Where instant communication with a caving party probably on the move is most missed, of course, is in flood warning.

One caving team did evolve a most ingenious method of automatic flood warning suitable for caves with an active stream running into the entrance and right through the system. When a certain amount of rain had fallen into a funnel, a small bottle of intensive colouring dye was tipped automatically into the stream, its bright coloration warning the underground team of possible trouble. Fine in theory, this idea did have the drawback of the sometimes considerable delay between the dyeing of the water and the time it actually reached the party. Another snag was that cave water often finds its way into public drinking supplies somewhere down hill and dale, and dyeing even for scientific purposes has to be undertaken with discretion. Still, the idea of the rain detector is a sound one, and coupled with a radio system which could send a warning signal to the team in the cave could prove a life-saver.

Whilst communications have advanced little in real terms for many years, the method of descending and ascending pitches is undergoing something of a major transition, particularly in Europe.

Although they still use lightweight electron ladders for the shorter pitches, American cavers – or spelunkers – have completely abandoned them for the really big shafts. Over the past dozen years or so, they have developed a method of tackling these which needs but one single rope, so eliminating the bulk and burdensome weight of the ladder.

The two main techniques now used in America, borrowed without any blushes from the climbing world, are abseiling (or roping-down) for the descent, and prusiking for the ascent. On reaching the top of a pitch, the caver belays his rope to a suitable anchorage point, throws the rest down the shaft, then abseils down like some great gangling spider on the end of its nylon thread.

The classic method of abseiling, as you will appreciate in a moment, has largely been superseded. This was favoured for its simplicity, for it involved no item of equipment other than the rope. This, the belayed end uppermost, was passed down between the caver's legs, around and up across one of the thighs, across the chest, then flung over the opposite shoulder to cross diagonally across the back to the controlling hand. Using the controlling hand as a brake and the other purely for balancing purposes on the rope above him, the caver could slide down the rope in a completely controlled manner, usually in a series of leaps. The friction of the rope around the body, plus the extra controlling grip of the hand, was enough to stop him plunging out of control. Anyone who has ever tried the classic method of abseiling down a rope will readily appreciate why mechanical devices were rapidly adopted by both climbers and cavers to act as friction brakes. While the pressure of the rope on the one shoulder was uncomfortably acceptable, the private damage being wrought by the tensioned rope through the groin was another matter. Bar room murmurs of emasculation put out of favour this 'one man and his rope' technique.

Such rope tactics have, of course, to be practised thoroughly on the surface before the caver ventures to use them underground, and even these apparently innocent daylight drops from branches of trees can have their dangers. Caver John Keefe decided that practice makes perfect, and slung a rope over a tree branch some thirty or so feet in the air. Having clambered to the end of the branch, he wrapped the old hemp rope around him in what he hoped was the correct configuration and launched himself. Almost immediately something went horribly wrong with what should have been a splendid demonstration of rope control. John turned upside down, but despite this graceless position – and a bright purple face – he managed to continue with a controlled descent. All might have been well, but within a few more feet the rough rope had burned through his thin shirt and began to tangle with the hirsute mass which he sported proudly on his chest. After a giddy moment of indecision, John realized that the only way down was – down. The spectators of the incident say that that last twenty feet was the longest abseil they have ever seen, and seem quite happy to recall in detail John's pronouncements as that hoary old rope denuded his chest hair by hair.

The American caver has undoubtedly taken subterranean rope techniques to their most sophisticated level yet. From the simple karabiner friction brake, he quickly evolved a system of small metal 'gates' latched across a special bent metal 'rack'. These gates can be added or removed to give greater or lesser friction, for while little friction is required at the top of a big pitch – the weight of the rope tending to pull the top section into a stiff rod of fibres – rather more is needed half-way down, and a great deal more as the abseiler nears the bottom.

Climbing back up the single rope demanded yet more development of techniques, and yet the system of jamming knots devised by Dr Prusik many years ago is still favoured above mechanical

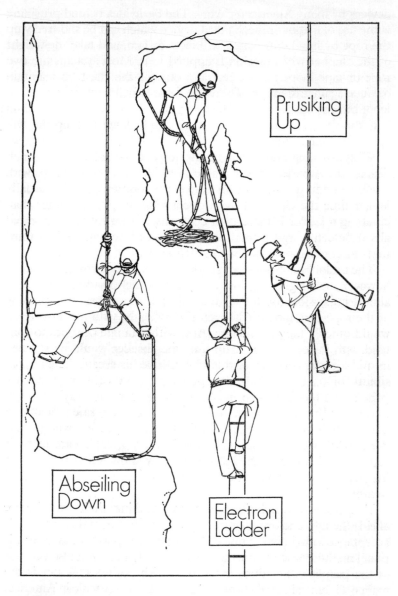

Vertical caving techniques

devices by many American cavers. The basic idea behind prusiking is the use of a knot or mechanical device which can be slid freely up the rope by hand, but which will lock, or jam, and take the weight of the climber when tension is applied to it. Most systems use two rope or tape loops for the feet and one for the chest, to maintain balance against the rope. There are some which have only one foot loop, but the climber has to be exceptionally fit to use this method – if the reader tries a one leg knee bend, down and up, he will appreciate why.

With abseiling and prusiking mastered to an astonishingly high degree, the American caver is able to tackle very deep systems with much less equipment than his European counterpart. The usually longer time the climber takes in prusiking up pitches rather than climbing a ladder is more than compensated for by the very rapid abseil descents, and a total tackle weight of perhaps only a third that of a conventional ladder and lifeline team.

The undoubted benefits of single rope techniques to caving must, however, be weighed against certain very real disadvantages. It is almost impossible to lifeline an abseiler or a prusiker; the spinning of the rope on a long pitch, even if very slight with modern ropes, would quickly tangle a lifeline. And with a tangle of ropes to contend with instead of one main line, the abseiler would be stopped in mid-air, the prusiker unable to continue his ascent. The impossibility of using a lifeline except perhaps on very short pitches (when even the Americans prefer to use ladders, anyway) has worried many European cavers trained to use the fail–safe system of a lifeline. It is worth noting, though, that those cavers who do use the single rope technique take considerably greater care of their ropes than do many of the ladderers. They know full well that their life depends on a single 'thread', and give their rope the cosseted treatment it deserves.

The American techniques are now making their way farther afield, though, and are favoured by many Continental clubs. Their acceptance in Britain is a much slower process, partly because traditional methods seem to die much more slowly here, but also because of the peculiar difficulties in many British potholes. Water is the main problem, a hazard encountered on a far lesser scale in America. The case is simply this: if a caver gets into difficulties on a ladder pitch, he can be hauled up or lowered back down on the lifeline.

When abseiling or, more important in this case, prusiking, he is utterly beyond help. As most dangerous ladder situations in Britain arise from exhaustion and exposure on a medium or long pitch – 70 feet and upwards – under a cascade of water, some extra security is obviously required.

Those British cavers who are using the new techniques argue that the caver will be far less tired on the ascent anyway, for the low weight of equipment and the abseiling of pitches requires much less energy output than when the same pothole is laddered. It seems safe to predict that the eventual outcome will be a hybrid approach, using single ropes on the long dry pitches and ladders on short or very wet ones.

Besides his basic equipment for descending shafts, the modern caver has at his disposal a number of other more esoteric items. These range from the winch and capstan, operated by brute man-power, electricity or petrol, to the Skyhook. This latter item – which takes its name from the mythical device which airforce apprentices are always sent to fetch from the stores – is a cunningly conceived device which enables the caver to tackle a pitch from the bottom, as long as there is an *in situ* rope passing up and through two steel eyebolts. Fixed to the top of the ladder, the skyhook is hoisted up the thin rope until it passes through the lower of the two eye-bolts and automatically locks in position. Such a device is used when a team makes frequent visits to a high level passage or aven (hole in the roof), but does not want to leave a ladder permanently fixed. A rearrangement of the skyhook by the last man down enables it to be released by a tug on the controlling line.

A natural question at this stage would be how are the two sky-hook eyebolts fixed in the first instance? This is where another gadget enters to enrich the caver's nomenclature, the maypole. Basically, the maypole is a number of sections of steel or aluminium alloy scaffolding tubes – of some six or more feet in length – which can be joined together by expansion or other types of joint.

The erection of a maypole is a clanging, clamorous process accompanied by a good deal of fervent praying. To the top of the first tube section is fixed the ladder, together with a pulley carrying a double lifeline. Another tube is jointed to the bottom of the first, another to the bottom of that, and so on until the whole contraption reaches the rim of the unexplored passage or hole high

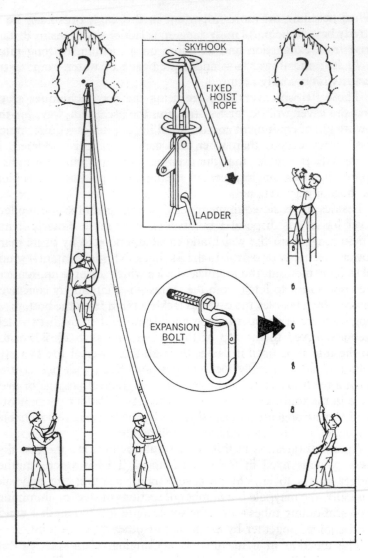

Maypoling techniques (*left*) enable cavers to reach unexplored high-level passages. Bolting techniques (*right*) serve the same purpose and can reach even greater heights. The upper inset shows a skyhook locking device for re-attaching a ladder to two fixed bolts above the ground. The man who has been doing most of the praying – he will be the lightest of the party and will have had no choice in the matter – then climbs the ladder, secured by the double lifeline from the bottom of the maypole. A climb of some thirty feet is usually without undue excitement, but much higher than that and the articulated pole begins to sag and sway like some swooning maiden. Guide ropes attached to the top section are kept tight to prevent the maypole from sliding out of place, but this only increases the load on the pole.

Obviously the techniques of shaft work are very important to a caver, but no less important are those which he needs to overcome the difficulties and dangers of the more or less horizontal stretches of passage. Here it is not a question of equipment, but of agility, balance and stamina.

There are four basic ways of making progress along a cave passage, each determined by the size and shape of the passage. A large passage is obviously the easiest of all, and except for the odd scramble over or around boulders can be negotiated by straightforward walking no more demanding than a mountain walk. A lowering of the roof to about four or five feet entails a backache-breeding stoop, whilst in a higher but very narrow rift the caver will have to shuffle sideways, and both are surprisingly tiring after a long distance. A very low roof forces the caver to his knees or even his stomach, and it is in this kind of situation that leg length is the critical factor. Where a caver of medium height can crawl on his hands and knees at a fair pace, back just scraping the roof, those with a slightly longer thigh will have to drop with a grunt to the most humbling of grovels. When the gap between the roof and floor is a matter of inches rather than feet, the caver resorts to the subtlest of his dark arts, that of squeezing.

For obvious reasons, it is the squeeze which most beginners fear in caving, and yet true squeezes are encountered far less frequently than straightforward flat-out crawls. Standard procedure is to enter the squeeze with one arm straight out in front, the other trailing by the body: this effectively reduces the shoulder area, the first of the anatomical problems except for those large-domed folk who claim 'If I can get my head through, the rest will follow.'

With his one arm straight ahead and the other pinned by his

body, the caver cannot rely on these limbs for much power, so he must push himself forward by his toes if holds are available for them, and also copy the wriggling and squirming motion of the earth-worm. The combination is sufficient to overcome the majority of squeezes, but the really tight ones – small enough to dictate that many cavers of broader girth can never pass through – demand even subtler movements. Negotiating the ultra-squeeze, pockets must be emptied of all objects liable to impede the caver's painful enough progress, and the chest can only be forced through by a

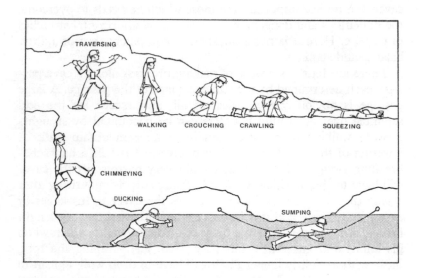

Progress through a cave

combined action of a deep exhalation and powerful pushing from the toes. Such a squeeze, which should really be left to the experts in extreme cases, demands a quite special frame of mind. Cavers know only too well that fear or panic can lead to a tensing of the muscles, with the result that the body is locked within the rock as securely as an expansion bolt. The ultra-squeeze needs complete relaxation, and infinite patience, with the body being allowed almost to flow through the constriction. Where the squeeze is some feet long, it may be many minutes before the caver can breathe properly, and he must know how to dispel the feeling of rising panic which can be triggered off by the pressure of the rock forcing him to breathe with a rapid panting.

The worst type of squeeze is that which is not only tight but zig-zagging, and it was in such a one that my own nerve was stretched more than on any other occasion. Our team was attempting to link the few remaining feet which separated two very long passages in Agen Allwedd, and I had spotted what I thought was a likely looking passage leading off from a small hole in the low roof. The passage turned out to be a squeeze, two feet high but very narrow, with right-angle bends every three or four feet. With my helmet and lamp pushed in front by my leading hand, it took only ten minutes or so to get to the end of the thirty foot squeeze. Here it petered out in an impenetrable fissure, with not even a suggestion of extra space in which to turn around. Going head first into a squeeze can be bad enough, but reversing through one has its own additional problems: one's clothing tends to be rucked up as far as the chest, the last place one needs extra inches in this situation, and of course there is no way of actually seeing the way out. My egress from this particular tube had gone with much difficulty until my feet reached the final acute corner – then came the big sweat.

With my head jammed against my now trailing arm, I was unable to see how to approach the bend, and had to simply follow my feet. But whenever I squirmed my feet backwards around the corner, both knees would jam against the wall in front of me. A full half-hour passed, much of it spent quite passively whilst I struggled in my mind to stop the feeling of panic. If I lost that particular struggle, I knew there was no chance of getting out, and none of the other team members were slim enough to negotiate even the entrance to that damnable hole.

I knew that this type of squeeze was like a lock, and that only the precise positioning of the body into the correct key shape allows one to pass through. Then I remembered the particular leg movement I had used on the way in: both feet, in their thick boots, had had to be lifted about a foot off the ground before the toes could pass through a small niche. With my thigh muscles cramping by now, I raised my feet time after time, toes groping blindly for that tiny channel which alone meant escape. On the tenth or eleventh attempt, with nerves at snapping point by this time, I found it. Exhausted, the key slid out of the lock.

That was some years ago, and only weeks before writing this chapter the Eldon Pothole Club made that connection of two passages – but not through that nightmarish squeeze!

Somewhere between horizontal and vertical caving techniques are those needed to traverse narrow rifts where the stream has carved a slot too narrow to negotiate at its base. Where a rift is wide enough, the caver can traverse with his hands and back pressing against one wall, feet against the other. When the rift narrows even further, the climber's chimneying tactics must be used, with the body jamming between the two walls in any convenient – if bruising – manner.

Some cavers like their caves dry and undemanding, but far more feel that the game has not really started until they are fighting their way down an underground torrent. The noise of the water varies from a contented murmuring in low languid stretches to a savage roaring of downward plunging white-capped waters.

The difficulties of travelling through subterranean water are not related to its speed or noise; the crashing ten-foot-high waterfall may be just an amusing challenge, while the slow black stretches might conceal submerged rifts or holes. Many cave rivers can be easily waded or swum, perhaps with the aid of a lifeline or tyre inner tube; others, like that in Padirac and other huge Continental river systems, require navigation in rubber dinghies. Steering such a dinghy through a narrow passage is a delicate and sometimes heart-stopping business as the paddlers push against the walls to stop their frail craft from puncturing on the many water-worn spikes and razor ridges.

Two final items complete our category of cave obstacles: ducks and sumps, again complying with the unwritten rule that caving terminology should be as baffling as possible to 'those outside'. The two are, in fact, slightly differing forms of water trap formed when the roof of the passage dips down to the water level. Where it almost meets, forcing the caver to duck completely under the water to get through, we have a duck; where it actually dips below the water level, a sump is formed.

The type of sump which is relevant to this chapter is the free-divable one – the more advanced and highly risky art of cave diving is covered later. With a sump of only a couple of feet in length the caver simply takes a deep breath and plunges through. Longer sumps usually contain non-rotting guidelines which the caver uses to guide himself through, secure in the knowledge that he will not find himself running low on his lungful of air in some blind cul-de-sac.

Until quite recently free cave dives were restricted to a maximum of perhaps fifteen feet, the rest being left to divers with breathing apparatus. But the last couple of years has seen more and more free dives of twenty and even thirty feet, where divers have already laid a guideline. A free sump dive of such duration calls for an extraordinary degree of self-confidence, for unless the free diver is being followed by a properly equipped diver, his chances of survival are nil should anything go wrong. And there is plenty to go wrong. Against medical advice, many free-sumpers still hyperventilate before taking the plunge – a method of taking many deep and rapid breaths to increase the oxygen level in the bloodstream. Unfortunately, hyperventilation is also a method used by researchers to induce fainting. . . .

Should he faint, knock himself out against the rock or become entangled in the guideline, the free-diver in a long sump has no air surface to float up to, no second chance. Only that single lungful of air to take him through the never-ending pipe of black water.

Such extreme free-diving has been condemned in some caving quarters, just as solo climbing has in rock-climbing circles. But both have taken their place as extreme extensions of the two sports, and are no less valid because of their extremeness.

Chapter 5

The Specialists

However sold on the idea of caving for caving's sake, the caver sooner or later finds himself having 'done' most of the technically demanding systems in his country and feels the need for something extra. When this stage is reached, he becomes a specialist in some branch of caving; this enables him to continue enjoying the sporting aspects of a trip down a cave, but in addition gives him some extra interest. Many turn to some branch of speleology, or cave science, but there are fortunately other less academic specialized pursuits for those of a non-scientific background.

Quite often this specialization comes about almost by accident. A caver particularly keen on finding new caves may adopt one of the newer cave detection or tracing techniques to further his quest, and before he knows it finds himself an expert in the particular technique.

Probably the most widely adopted cave science is the study of how caves were formed, and still are being formed. The popularity of this particular form of study is easily understood, for there is no caver who at some point in a trip has not stopped to wonder exactly how that particular section of passage or shaft was formed.

In many respects, the study of cave formation is the easiest and most convenient of the speleological sciences to follow. To understand the whole picture of a cave system one simply follows it to the end – or the bottom in the case of a pothole – analysing the various stages and processes of development as you go along, so enjoying the sport provided by the cave at the same time. But although it may be the most obvious specialization to adopt, it is probably the one most full of doubt. Many are the wrangles and arguments over the precise processes which led to the making of a particular passage, or even a whole cave. An amateur in the subject, or even an expert, may have covered two-thirds of a cave and formed a pretty sound idea as to the processes which formed it – and then find a section of passage which knocks the theory completely off its pedestal.

Obtaining anything like an accurate picture of how a cave was formed might take an expert many years of devoted study. Not all his work will be underground, either, for he has to take full account of the surface landform plus effects of varying rainfall over many thousands of years, the ice ages, and the fluctuation of local water tables.

Allied closely with the sporting caver's interest in underground water is the spleleologist's concern with hydrology. If limestone rock is the mother of caves, then water is surely the father. There are two aspects to the study of cave water: the make-up and behaviour of water in cave formation, and its vital importance as an indicator of unexplored cave.

The first part, the study of cave water for its own sake, is self-evident. Speleohydrologists will measure its temperature, hardness and rate of flow to determine the part it has played and still has to play in the manufacturing of cave passage. But even more intriguing is the rôle of cave water as detective in the search for new cave.

This business of seeking new cave (a handy phrase which covers both entirely new cave systems and straightforward passage extensions in known ones) is a dynamic driving force within the sport and its associated science. There is a simple test to apply to anyone who claims to be a true caver: mention in the most casual manner the existence of some unnoticed and unexplored tube within a known cave, adding, perhaps, that you really must have a look in it at some convenient time or other. The pseudo-caver will merely wish you good luck, but the true representative of the race will assume a glazed expression and try to appear rather more interested in the weather as he plies you with polite questions as to the exact whereabouts of this tube. And he will also wish you good luck with it, naturally.

Opportunities for poaching new cave are, however, relatively rare. Any caver or club team with a promising find in their files will usually treat it as a high security risk until they have made the breakthrough and surveyed the discovery. The exception to this is the very large – or very obvious – cave dig where extra outside help is welcomed.

To our subterranean detective, cave water provides the vital 'fingerprint' of unexplored passages. He knows that the water entering a live or active system must re-emerge somewhere, and that somewhere will be dictated by the local geology, water table, and other factors. He must be skilled at interpreting maps, mating the knowledge they provide with his own knowledge of the cave passages explored so far; if his deductions are correct, the cave seeker can trace a thin red line over the map to the resurgence and state, 'That's new cave . . . let's get working.' Often the job of pushing known passages within a cave is out of the question, and the only hope of continuation is by looking for a downstream – and possibly choked – entrance.

Before a team commits itself to many long hours of work on a possible extension, the hypothetical connection will have to be proved. It may seem obvious that the water going into cave X will emerge from point Y, but many such 'safe' proposals have been proved wildly incorrect after a water tracing test.

Water tracing is one of the more colourful aspects of speleological research, as anyone who has seen a few ounces of the dye fluorescein dissolved in a stream will testify. Although non-toxic, this dye is one of the most powerful known and can be detected by the naked eye even when diluted to one part in ten million. One pound of fluorescein is sufficient to treat ten thousand million gallons of water, and by the use of special detectors (which can be left lying in outlet rivers) and observation under ultra-violet light, even greater dilutions can yield a positive result. Early experiments with this dye taught water tracers to proceed with caution, though, for an over-generous application of fluorescein to a cave river or stream may easily result in somebody's drinking water supply suddenly turning a brilliant green. Assurances as to its non-toxicity do little to assuage tempers when this happens, and one group of French cavers found themselves being chased out of a village when the main fountain – the only source of drinking water – turned Technicolor overnight.

Other tracers which are used include common salt and minute lycopodium spores. Salt seems an attractive alternative, as it gives no colour to the water and can be detected by a simple resistivity meter, but it needs a ton of salt to treat the amount of water which can be treated with just a couple of pounds of dye. Sometimes the water tracers have their positive results within a matter of hours, but there are other times when they need great patience. Fluorescein added to the water in the huge French river cave of Padirac took over three months to travel seven miles, and half a hundredweight of dye poured into la Henne-Morte – sufficient to treat half a billion gallons – was never seen again, lost in what the French described as the cave's 'vast digestive tube . . .'

Besides using the study of underground waters for tracing unexplored caves, it has also very important applications for society as a whole – so much so that when a caver is asked 'But what possible use is caving?', he can cite a number of cases where speleological research has made a definite contribution to the public good. The main by-product is, of course, the location of sources of drinking water, and with the world's demand for fresh water rising inexorably this will be a more and more sought after service. Only cavers can tell whether a cave system is capable of sustaining a regular water output of a certain velocity, important when the water takes a certain minimum time underground before all impurities are settled out or decomposed. And only cavers can trace possible sources of contamination, such as the many open shafts used by farmers for the dumping of dead cattle and rubbish. Some of France's most ambitious hydro-electric schemes came about after cavers discovered vast and reliable water courses under the mountains. When the flow of the Nile was checked by the Aswan Dam in Egypt, the water did not rise as high as predicted: too little attention had been paid to the many small caves in the Nile valley, and these were draining much of the water through to the Red Sea.

One of the strangest hydrological problems ever tackled by cave scientists was the case of the water mill on the coast of the Ionian island, Kefallinia. A strange mill, for the water which drove its enormous wheel was sea water, flowing inland! Just as Alice in Wonderlandish was the fact that this water re-emerged, brackish rather than salty, three feet above sea-level on the other side of the island. Theories that the sea was leaning here did not quite satisfy the backroom boys, and the findings of a team of Austrian cave divers were eagerly awaited. By diving and extensive dye tests – using up to 270 pounds of fluorescein – they found the answer to the riddle: the sea water plunged deep into the caves and there mixed with fresh water coming in from caves high in the mountains.

This made the water, now brackish rather than salt-laden, less dense so allowing it to rise up on the farther side of the island by the few extra feet necessary to let it run down to the sea like any self-respecting stream. Without the help of cavers, this 24-mile-long intercepted U-tube would have remained unknown, and mystifying.

The popular impression of a cave is that of a barren hole, devoid of life except for the shufflings and scramblings of visiting explorers. But with few exceptions, caves contain their own forms of flora and fauna, tightly knit and utterly interdependent communities adapted to the realm of perpetual night. The study of underground fungi and creatures is an absorbing and fruitful one, free

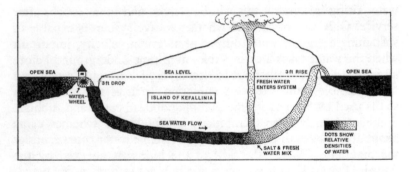

The secret of the Kefallinia waterwheel

to a very great extent of the effects of pollution, pesticides and other products of the human race's efforts on the Earth's surface. Speleological research is still in its infancy compared with other earth sciences, and the caving scientist stands a better chance than his sun-kissed brethren of making a unique discovery.

Although the first permanent cave dweller to be described was the cave salamander, Proteus, in 1689, biospeleology did not really come about until the beginning of this century; it was not until 1957 that the existence of a true cave fauna was accepted.

One of the most remarkable things about natural life underground is the finely balanced food cycle, often approaching a perfect closed ecological system – but not quite, for the sun's energy must be utilized somewhere along the line, however indirectly. Material brought in from the outside provides the chief initial supply of sustenance, but there are also certain bacteria which live simply off the minerals in mud deposits.

Most cave life is far from obvious to the casual observer, who must get down on his hands and knees and peer at every pool and stretch of rock or mud before seeing the tiny betraying movements of a cave insect. Besides the European blind and colourless Proteus salamander, which owes its survival to its ability to live without eating for years at a time, other more readily visible inhabitants are the cave fishes of America and Central Africa. American farmers are often surprised to find a colourless crayfish in a freshly pulled bucket of well water, the creature having made its way along the underground watercourse.

Snakes dislike the high humidity deep inside a cave's dark zone, but some of the largest collections of rattlesnakes in Texas and Oklahoma are housed in dry cave entrances, and Malaya has its own bat-eating snake, a Golum-like horror which wraps itself round any sleeping bats it comes across to crush the life out of them.

The subject of bats was bound to come up at some point in a book of this nature, and the very word is probably enough to have sent shudders down the spine of many readers at the first sight of it. Years of Dracula films and stories have established every species of bat as blood-sucking vampires, and other tales of their 'deliberate' attempts to entangle themselves in women's hair have not helped to get the true picture through. True, there are some Mexican cave-dwelling bats which do drink the blood of cattle and horses, having slashed them – unnoticed – with their razor sharp teeth, but the majority of bats are quite innocuous, except as unwitting carriers of rabies.

Bat researchers have discovered a great deal about these remarkable creatures, such as their echo-sounding techniques for flight guidance in narrow passages and detection of insects on the wing, but there is still much work to be done in this field.

The bat colonies of some European caves reach quite a respectable size, but the bat caves of Texas often house populations running into the millions. The nightly emergence of so many bats from one single entrance is an unbelievable sight, and legend has it that this was responsible for the discovery of the huge Carlsbad Cavern. A young cowboy by the name of Jim White could not believe his eyes when what he thought was a huge cloud of smoke turned out to be bats emerging from the unexplored cave at the rate of one hundred every second. His early explorations, alone with a single paraffin lamp, led to the discovery of some of the greatest chambers in the world, but only gained after he had overcome the huge piles of bat excreta known, like that of sea birds, as guano. Some Texas bat caves have deposits of guano fifty or more feet thick, important sources of fertilizer and even nitrate for explosives in the American Civil War. Each bat contributes less than one ounce of guano to the common pile each week, but when multiplied by some millions a year's output can be compared favourably with that of an efficient factory.

British researchers can handle their bats with impunity, but the discovery of rabies in American and some European bats has led to rather more cautious handling by cavers in these countries. Two non-cavers died from rabies after visiting the giant bat home of Frio Cave in Texas, neither having been bitten. Experiments were conducted with caged animals left in the cave, and when these contracted rabies even after isolation from everything but the cave atmosphere it seemed the air itself of large bat caves could transmit the disease. American cavers console themselves with the thought that, as yet, no caver has contracted this dreadful disease from a bat bite.

There is a quite reliable rule-of-the-thumb method for differentiating between a 'bug man' and a 'bat man': while the bug man ambles through a cave with his head bowed, as though in constant shame, the bat man gazes intently at every inch of the roof, apparently expecting an imminent religious experience. Whatever his attitude, the student of life deep in the rock finds his own special rewards. He may labour for years in the cataloguing of various standard species, but he knows well that the very next trip might bring some completely new find into the glow of his lamp. One particular dream of speleomicrobiologists is that they may share the luck of Fleming and stumble upon some completely new and powerful antibiotic. After all, only now do we know that the therapeutic properties of moonmilk – an odd soft white material found on some cave walls – claimed by sixteenth- and seventeenth-century physicians was due to the presence of anti-biotic actinomycetes.

Two other fields of specialist cave research deal with cave life long dead – palaeontology and archaeology. As mentioned before, caves can act as time-locked treasure houses storing the remains of the men and animals that roamed the Earth thousands of years ago. Active streamways hold their secrets only briefly, until the next flood, but the drier parts of caves which were once readily accessible from the surface may hold the information missing from our books on prehistory.

One of the most remarkable subterranean prehistoric discoveries was made by the indomitable Norbert Casteret way back in 1922. Completely alone, he dived unclothed into an unexplored sump in the Montespan Grotto, France, his candle and precious matches stowed inside his bathing cap. Beyond the black sump he discovered priceless drawings and clay models of now extinct animals.

Not every caver looking for a speciality has a scientific turn of mind, though, and many turn to a more technical aspect. Of these interests, photography is by far the most popular.

Many budding cave photographers start off with grand ideas of providing a complete picture coverage of a sporting trip, but such notions are quickly dispelled. Photography underground presents so many problems, even to the expert, that it is very difficult indeed to photograph a team of cavers on a normal trip; special photographic trips must be arranged, with the muddy models prepared to stand around for hours on end. This is asking a lot of them to start with, but volunteers for these trips dwindle in number when the photographer makes more and more demands of them. The discerning cave photographer will not be happy until his subjects are standing up to their necks in some freezing river, or fighting their way up an equally cold but even more spectacular waterfall. The first two flashbulbs will refuse to function (except until the moment of withdrawal as the cameraman tears them out in disgust), and the models will find themselves holding quite impossible positions for anything up to half an hour. Such treatment leads them quite quickly to the conclusion that the photographer is a ten-thumbed bumbling incompetent, an unfair judgement considering the multitude of difficulties facing him.

The surface photographer slings a camera round his neck and

wanders about at will, taking his pictures wherever and whenever he wants. His only concerns are subject framing, light availability and motion of subject – and hardly the latter two with modern automatic cameras. His underground counterpart needs almost superhuman patience and perseverance. First, the problem of getting the equipment undamaged to the chosen site within the cave. Favoured by most is the pressed steel ex-service ammunition box which, with an extra watertight seal around the lid and internal foam rubber padding, will keep the camera and other equipment both dry and free from most shocks.

The second problem comes when the photographer, hot from the exertions of dragging his precious tin box through the entrance passages, comes to unload the equipment. Cavers' hands can hardly be described as surgically clean half-way through a trip, so the photographer must wear gloves when moving and swab his hands down with a towel as soon as he opens the box. The removal of the camera and the setting up of the tripod and lighting seems an unnecessarily drawn out affair to the onlooker, but the experienced cave photographer knows from bitter experience that it will take a long time before the camera lens demists after its transition from the dry box into near saturation point humidity.

The absence of any natural light poses some problems, but in other respects gives the cave photographer a far greater freedom. With good flash bulb or electronic flash sources of light he is able to sculpt his subject much more accurately than on the surface, but always with the proviso that the lighting *is* good. The cave environment can play havoc with electrical connections, however hard the photographer has tried to keep them clean, and a tiny piece of dirt infiltrated into a socket can take a long time to locate.

Really large caverns need quite enormous amounts of light, and the two ways in which this can be provided are by using flash bulbs – as in the now classic National Geographic photographs of the Big Room in America's Carlsbad Caverns when many thousands were simultaneously fired to provide a massive pulse of light – or flash powder. Cave photographers seem to get their own special version of the seven-year itch, and find the challenge of photographing a big chamber irresistible. Unable to afford thousands of flash bulbs, they turn to flash powder.

A standard question now asked by caving models before a camera trip is: 'You're not going to use flash powder, are you?' A deliberately rhetorical question, for they know the stupefying effects this substance has when used in a cave. Even the allegedly 'smokeless' varieties emit such a dense choking cloud of foul fumes that the chamber has to be evacuated instantly. One such event every seven years is enough to satiate even the most ambitious photographer's appetite.

Despite the drawbacks already mentioned, plus the ultimate heartache of finding rolls of developed film quite blank or grossly under- or over-exposed, the cave photographer continues happy in the knowledge that his few fine prints or slides are unique recordings of some of the most remarkable scenery anywhere. And one cannot but admire his perseverance, for caves take their toll of cameras with expensive regularity. Ask my team-mate Bob Fish about the drowning of his Pentax, but gently.

Even more technically demanding is cinematography underground. While the cameras can be protected by the methods described above, reliable and consistent lighting is required for minutes at a time. Recent developments in battery design have reduced their size considerably, but cost often prohibits the use of the newer but very expensive power sources. A general film about caving can be made by running power cables in from the surface and confining the shooting to within a few hundred yards of the entrance, but this is obviously not possible when the film is intended to portray a trip down one particular cave of some length. Nor can petrol or diesel generators be used underground where adequate ventilation can certainly not be guaranteed. The usual answer is to manhaul car or lorry batteries into the cave on special backpacks, taking them out when recharging is necessary.

The really first-class caving films in existence could probably be counted on the fingers of one hand, but the standard is gradually being raised as cine photographers, of the standard of Pierre St Martin veteran Sid Peroun, realize that only their work can put across the full graphic glory of caves and the problems of reaching into their innermost depths. There have as yet been few television screenings of caving films – mainly because of the scarcity of really good ones – but at least cavers have made a small dent in the generally bad opinion of them expressed by the public.

While photographs and films provide a general idea of a cave's layout, the surveyor is the man responsible for measuring and drafting every twist and turn and convolution of every system, from the simplest to the deepest and most difficult in the world.

The rewards of cave surveying are twofold: there is the pleasure of seeing the cave's form taking two-dimensional shape on a hitherto blank sheet of paper when the scrawled notes are deciphered from the muddy notebook, and there is the actual task of surveying itself. To survey a cave properly – 'Is there any other way?' the fanatics will cry – is a task which takes the surveying team into every nook and crevice, with not only time but a definite necessity to look around each section of passage. Besides giving the surveyor an unparalleled knowledge of the cave, it also gives him plenty of opportunities for fresh discoveries. That low muddy bedding plane leading off from a main passage may have been ignored by all other parties, but not the surveyor. Dragging his tape and compass through, he may well find a new extension – and the job of surveying something you have just found yourself is a doubly enjoyable one.

The actual techniques used by a cave surveyor vary according to circumstances. The old Cave Research Group of Great Britain graded surveys in accuracy from I to VII, numbers which denote surveys from rough sketches drawn from memory to ultra-high grade ones using sophisticated mine survey equipment. The uses of surveys are to acquaint cavers with new systems and to indicate general trends of the cave development. A further vital use – not yet drawn upon – is the possible need for a high grade survey to enable a rescue shaft or tunnel to be dug from the surface if a party were trapped by a roof collapse, or if such a method of evacuation was judged best for a badly injured caver.

There are many other branches of cave study, from research into the subterranean atmosphere to the study of cave formations, but the other main specialization is digging.

Digging in caves needs considerable dedication, an utter disregard for discomfort, and nerves of steel. The dedication comes in when a dig has been worked on for weekend after weekend for perhaps one or two years without results. The mastermind behind the particular project finds it increasingly difficult to muster manpower for the backbreaking work, and is fully aware that if his club drops the idea some other team may well come along and make the big breakthrough in just one more session. In fact, cave digging is the most difficult pursuit in which to call a halt; when you have gone so far, with so much labour, it seems a tragedy to drop it, and in the early stages one feels that it is far too soon to abandon hope. A pretty dilemma!

This is the era of the cave digger, for by his work – and that of cave divers, examined later –   come many of today's new discoveries. Except for the vast speleologically undeveloped regions of the world, there are few open entrances unexplored, few obvious and accessible passages still virgin. But there are countless thousands of opportunities for cave digging, where progress is halted at the moment by silt, mud, boulders, or a mixture of all three.

The tools at the digger's disposal range from the humble trowel (often the only implement which can be wielded in a really tight space) to power drills, winches, railway tracks for spoil disposal, and explosives. The latter, usually in the shape of gelignite or blasting gelatine, is particularly useful when solid rock is encountered, and saves days of hammer and chisel work. Naturally, the use of explosives – normally called 'bang' by the caving fraternity, or coyly referred to in written reports as 'chemical persuasion' – is rigidly controlled in most countries, and in Britain at least, the handler must be licensed by the police. Although there have been no accidents caused directly by an explosion in a cave, to my knowledge, gelignite fumes have sometimes been a problem in draught-free digs. The effects of these fumes can take some time before they are obvious, as one South Wales caver found. He spent the evening after a banging trip in the pub, apparently quite healthy, but came close to death later that night when the fumes had been fully assimilated by his body. It was only by chance that his unconscious body was found during the night, half out of his bunk in a desperate attempt to go for help, and he was rushed to hospital.

Underground digs are usually quite obvious in their siting, but some surface digs start at an apparently random point on some featureless plateau or moor. These may will be the result of cave detection by electrical resistivity tests: a series of metal pegs are stuck into the ground in a straight line, an electric current is passed through the ground, and the resistance met along the current's underground path is carefully measured. Where a cavity lies in the path of this current, much higher resistance is met, and careful interpretation of the meter readings can give some idea of the cavity's size – whether it is in fact just a crack or a full-size cave.

Cavers may be prepared to dig a shaft down to a promising cave predicted by this method, but few will be keen to spend the same time on a project proposed by that most scarce of cave specialists – the diviner. With his little curves of twig or metal, the cave diviner is on a thankless task today, but with the caver's greed for fresh cave, who knows what we will be believing in tomorrow?

5. Moorwood Sough, a mine drainage level at Stoney Middleton, Derbyshire. Exploration of such disused mines is yet another specialised aspect of caving

Low, wet and sharp-rocked crawling – so much a part of modern British caving. This crawl leads to the Bony Hole series of Knotlow Caverns, Derbyshire

6. The tenacity and all-pervading nature of cave mud is quite remarkable, as witnessed by the generous coating on this caver about to abseil into a chamber in Dr Jackson's Cave, Derbyshire

Chapter 6

The Cave Divers

Of all the branches of caving, there is one which is set apart as something rather special, almost sacred – at least to those cavers who observe rather than take part in it. Despite the many sumps to be found in British caves, the really experienced cave divers of this notoriously wet country could be counted on the fingers of three, perhaps four hands.

Cave diving has developed as an activity for the very few, the élite. But it must be put on record that the air of mystery which seems to surround cave diving has been created by ordinary cavers, not the divers. To the caver whose sub-aquatic activity is limited to holding his breath and diving for a few feet through a short sump, the epic ventures of those air-tanked, masked, lead-weighted and flippered few seem completely beyond his ambitions. When pub talk amongst cavers turns to cave diving, a definite degree of respect almost reverence – is manifested.

Cave divers would themselves be the first to ridicule the idea that they are in any way 'special'. They maintain that they are simply cavers with a rather specialized interest – which is true. And most will even maintain that diving is by no means as risky as is commonly supposed – which is doubtful.

Why, then, do common-or-garden cavers persist in regarding divers as the cream of their kind? I think it is because a cave diver has no secondary line of defence; he is, to all intents and purposes in modern diving, utterly and completely on his own. He relies for each gasp of precious air on a conglomeration of hardware with tiny valves and delicate springs in amongst which a single grain of sand can wreak havoc, and when you are slowly finning your way along a completely submerged tube, perhaps hundreds of feet long, the last thing you want is havoc.

This is the nub of a caver's deep respect for divers: that placing of one's trust in a relatively complex and therefore vulnerable piece of equipment, and this in a sport where even simple items such as ropes and ladders are always regarded with a small but persistent level of distrust. Whether one likes wet caves or not, no one likes the idea of a completely flooded passage – especially with you in the middle of it. This is the stuff that post-big-trip nightmares are made of, the shallow-sleep struggles as flailing limbs become entangled in the bedsheets and the floods of your dreams close over your sweat-covered head. Cave divers have no time for such fantasies: they don their gear and get on with the job of turning a nightmarish situation into just another caving trip.

Cave diving started in Britain in 1934 when a young northern caver, Grahame Balcombe, assembled a weird and wonderful piece of apparatus on a convenient area of rock upstream of Sump One in Swildon's Hole. Swildon's has for long been the most popular cave in the Mendips, and the full story of its slow but steady exploration stretches from the first trip in 1901 to the present day. Much of the effort in exploration has been centred on the passing of the system's many sumps, so it is fitting that the cave should also have been the cradle of cave diving.

Divers of today would shudder at the sight of Balcombe's rudimentary equipment – but then innovators never do have the benefit of hindsight. Despite its Heath Robinson appearance, Balcombe's diving outfit was ingeniously contrived, consisting essentially of a valve unit based on an old bicycle frame, a few tin cans and 40 feet of garden hose. Clasping the mouthpiece between his lips, Balcombe lowered himself into the water – with what degree of confidence I cannot tell – and pushed his head into the shallow sump. The equipment had only one flaw: it did not work, and Balcombe found himself unable to draw any air through it when only a few inches down. His companion, Wyndham Harris, had an equally unsuccessful try, then Balcombe again. With his arms held rigidly in front, the plucky caver was convinced he felt an air space. He was also convinced of an obstruction on the far side, and hastily retreated. Sump One had held on to its secrets.

A couple of attempts at blasting Sump One out of existence with large amounts of explosives failed in their desired effects, but did arouse antagonism amongst local inhabitants when their sleep was shattered on one occasion and their Sunday prayers in Priddy

Church on another. Brute force having failed, science was again resorted to, in 1936. This time the equipment was rather more sophisticated: a waterproof suit, a face-mask, and a small hand-pump to supply air to the diver. Sadly, Balcombe was not able to go, but another companion of his, Jack Sheppard, strapped himself into the baggy garments and successfully passed through the sump.

There is one puzzling tailpiece to the story of Sump One in Swildon's. Since that very first passing in 1936, the length of the sump has always been described as between six and eight feet. Literally thousands of cavers have free-dived it over the years, and never a word of dissent over the standard description . . . until the University of Bristol Spelaeological Society turned to science again, in 1970, and ran a tape measure through it. Two feet and eight miserable inches. Such is the breaking of a legend.

Whilst Swildon's Hole was the scene for the first real cave diving attempt, that by Balcombe, and subsequently witnessed many first-rate diving exploits, it was in another Mendip cave that the sport was really developed over the years to its present level. That cave is Wookey Hole, known – and used – by man for thousands of years. The Fourth and Fifth Chambers of Wookey were once easily accessible to visitors, but then a small paper mill was built just downstream of the cave entrance and a dam was built, flooding the cave so that only the first three chambers could be reached.

Today, Wookey Hole is one of the most popular tourist caves in Britain, with well over 200,000 visitors a year treading its concrete paths and admiring such formations as the Witch of Wookey with the aid of concealed lighting. When Wookey Hole came up for sale in early 1973, the price paid for the whole concern was £400,000, evidence indeed of the cave's commercial value. Madame Tussaud's, the world-famous waxworks company, who purchased Wookey Hole, have quite rapidly followed through their ambitious plans for the blasting of 500 feet of artificial passages to enable the public to see the wonders of the Ninth Chamber. This must be the first time that the owners of a business have been prepared to invest so much money to open up a tourist attraction which they themselves have never seen, except on photographs taken by cave divers. With the extra chamber opened, and an artificial passage blasted to the cliff face looming over Wookey's impressive river entrance, a round trip will be possible for the first time, and Madame Tussaud's hope eventually to attract in the region of nearly half a million visitors a year. To the owners of Wookey Hole, then, cave diving has been far more than something to be tolerated.

In 1935, the year after Balcombe had made his first spluttering 'attack' on the sump in Swildon's, there began a carefully planned and energetic assault on the sumps of Wookey Hole. There were six divers: Balcombe, the leader, Wyndham 'Digger' Harris, Frank Frost, Bill Bufton, Bill Tucknott, and – shades of Women's Liberation even in those days – Phyllis 'Mossy' Powell. The whole exploit was made possible by the generous donation of equipment by the diving company Siebe, Gorman & Company.

Those early dives revealed many of the persistent problems of this form of exploration: difficulty in communication, dive after dive required to make even the smallest headway in difficult terrain (or should it be aquain?), and appallingly low visibility. This latter problem is often the greatest faced by British divers. American, South African or European sumps are often on a vast scale, with the divers well clear of the sides and floor. In Britain, sumps are, as often as not, constricted places with liberal deposits of silt and mud. However carefully the diver enters the water, even with modern streamlined equipment, the slightest disturbance stirs up the sediment and visibility rapidly worsens until he quite literally cannot see his hand in front of his face.

Balcombe and company countered the visibility problem by using a technique still adopted today, the laying of guidelines which are left *in situ* for divers to follow. The progress they made in 1935 was quite remarkable, especially when one considers that the diving equipment was of the sea-diver's hard hat variety, with the luckless individual inside his brass dome having to drag hundreds of feet of air pipe after him. The divers had reached the Sixth Chamber when increasing complaints from the villagers of Wookey Hole forced the team to prepare for one final now-or-never push. The complaints were about the state of the drinking water on mornings after the divers had been down, stirring up the mud despite every precaution.

Balcombe dived at just before midnight on August 31st. One diver had previously reported what he thought was an air surface, which would be Wookey Seven, but would they be able to reach it on the final attempt? 'Mossy' Powell was Balcombe's back-up diver and all was going well until, just as Balcombe was about to pull himself up into the new chamber, Mossy's wrist seal came undone. This had been a recurring snag, owing to her small wrists, and the two had to return to base. The leak was repaired, and the two divers set off again. Balcombe's diver's log tells the rest of the tale . . .

'00·52 Re-entered water. 00·53 Under arch. 00·54 Fouled light cable to Fourth. Clear. 01·06 In Six, relaying shot 4. At Belay 1. Searching for iron weight. Found weight. 01·10 Trapeze anchored. 01·14 Depth 11ft. 01·19 Drums full. 01·20 Trapeze at surface. 01·24 Pumps running at 60. 01·27 No2 (Mossy Powell) approaches. 01·36 No2 re-belayed. Slack pulled down to No1. 01·40 At the surface on trapeze. 01·43 Pump steadily. (That wasn't what he said!) 01·46 Difficulty in obtaining proper field of vision at the surface. Chamber so high that the top invisible. Likewise end not visible. Breadth about 20ft. No appreciable stalagmite formation. Rock still conglomerate.'

Balcombe had made it. He was through into Wookey Seven, and the last trip of the year was crowned with glory. This is how Balcombe described the Seventh Chamber: 'The Great Seventh, though only some twenty five feet wide, transcends in architectural magnificence all the other chambers of the cave, rising with sheer and parallel walls, straight up into the blackness impenetrable by either the 100-watt lamp or the diver's torch, and running as a mighty rift, away into the unknown, the end, like the roof, swallowed in the gloom.'

The 'trapeze' mentioned in the diving report was a float which was sent up to the surface trailing a rope, and up this the diver could pull himself to the chamber.

So it started, and so the Wookey Hole story continues as dive after dive pushes the known limits of the cave closer and yet closer to the sink at Priddy Green where it all starts. At the time of writing a handful of crack divers have pushed exploration as far as the Twenty-second Chamber, and as the dives get longer, so the logistics become more complex, and the problem of adequate air supplies more acute. Rather as mountaineers have to build up a chain of camps to enable one or two to reach a summit, so cave divers nowadays must often take spare air cylinders to depot along the way for the return journey, or emergency use if the dive is likely to use most of the air they are carrying.

The end of the war saw a renewed interest in cave diving in Britain and – most important to its future development – the formation of the Cave Diving Group. The development of frogmen's equipment and techniques during the war gave the divers new weapons, principally the closed-circuit oxygen rebreathing apparatus. The next 15 years saw nearly all cave dives being made on this type of kit, in which oxygen is fed as required to a bag on the chest of the diver who breathes directly from this bag. Carbon dioxide is absorbed by a cylinder of soda lime. Whilst oxygen equipment is small and has a long duration, the hazard of oxygen narcosis prevents its use below a depth of 33 feet, and it is really too complex for cave work. But it did provide divers with the first opportunity of working independently without the limitations of the hard-hat suit.

During the years in which Wookey Hole was the centre of the CDG's attention, some weird and wonderful items of equipment were developed and eagerly copied by others. The 'Aflolaun' was the most remarkable of these, and an Apparatus for Laying Out Line and Underwater Navigation became *de rigueur*. The 'Aflo' usually contained a line reel, car headlamp and battery, compass, depth gauge, signalling horn, watch, thermometer, telephone connection and writing pad, together with a few spares to keep the whole thing functioning. Whilst pushing an Aflo through the fairly open sumps of Wookey might not have been too much trouble, sumps in other caves such as Swildon's Hole or Stoke Lane Slocker were not as accommodating, and it is murmured in some quarters that the Aflo probably held up the progress of the sport rather than assisting it. Cave divers of the 'seventies have a quite different attitude, and keep their ancillary items down to a bare minimum.

The most significant change in cave diving occurred in 1962 when Mike Boon proposed an air breathing set suitable for cave conditions. His argument was clinched when he passed Sumps Six and Seven in Swildon's, the latter by pushing his air bottle in front of him – something an oxygen-kitted diver could not do. Since then, British cave diving has been carried out almost entirely on air. The bottles are larger (now usually strapped to the diver's side to enable him to pass through constricted sumps), but the equipment is less complex, and therefore less vulnerable, than that based on oxygen. With a 'floor' of 300 feet, air-breathing divers do not yet have any problem over deeper sumps, either. With the advent of the wetsuit, gone too are the problems of having to guard against the puncturing of the early dry-suits.

Something else prompted the eventual switch to air: the death of James Marriott in Wookey Hole. In April 1949, Marriott joined a team of CDG divers in a push through to the Ninth Chamber. Although not a member of the Group, he was a trained frogman and knew his oxygen equipment well. Despite two of the six divers having to return to base due to facemask trouble (one of whom was the pioneer Balcombe), the other four reached the Ninth Chamber without bother. Two went through to the Eleventh Chamber and then returned to the Ninth. Because this was to be his only trip in the cave for some time, Marriott asked his companion, Dr Bob Davies, if they could exit via a longer route than that which the other two were to take. The two parties split, and, having dived with him to the Tenth Chamber, Davies showed Marriott the correct guidewire to follow.

With Marriott on his way out, and the twitching of the guidewire to show that he was on the right route, Davies dived last. He continued right through to the base in the Third Chamber, but was horrified to hear that Marriott had not emerged. Somewhere in the water, with visibility down to near zero, he must have passed him. Davies' own oxygen was at a dangerously low level, but he dived back into the sumps immediately.

Davies swam until his oxygen gave out, then switched on his limited emergency supply. At this moment another of the divers, Donald Coase, passed him. Fifteen feet farther in, Coase found Marriott's body, his cylinder and breathing bag both exhausted.

Later examination showed that a test gauge used by Marriott was faulty, leading to a far higher flow of oxygen than was necessary, or safe. Marriott's life had literally wasted away with each bubble of unused gas.

Discussing diving techniques with Mike Boon, I asked him if air had replaced oxygen for good: 'No,' he replied, 'I think that in some cases the wheel will come full circle. Mike Wooding, for instance, has been doing some very long dives indeed in Yorkshire – over 1,100 feet in Keld Head, for instance, which must be one of the longest cave dives ever done in Europe if not in the world. Keld Head is a very long but fairly shallow dive, so an oxygen kit of high duration could well be an advantage there. Another possibility in these long dives is the use of an oxygen/nitrogen mixture which enables you to go deeper than on pure oxygen alone. The real problem is that you have to be a very experienced and well-trained diver to use oxygen or mixtures. Mixtures, particularly, are very complex, and need very accurate flow settings.'

A helium mixture was used in the late 'sixties by a British diver, Frank Salt, to a depth of 355 feet in Rhodesia's Sinoia Caves, but the passages were huge, the water had a temperature of 72°F and gave 200 foot visibility.

The actual method of progress through a sump has two quite separate followings. There are those who prefer to bottom-walk and those who prefer the speedier process of actually swimming with the aid of fins. I asked Boon which he preferred.

'I would rather bottom-walk, even though some of the best divers today are finning. Finning is faster, but you're dealing with three dimensions then instead of two. Also, if you hit something, you'll hit it a lot faster if you're finning. Mark you, if you want to get out of a sump in a hurry, fins are obviously a damn' sight better than trudging along the bed of a sump.'

There is a deeper and more serious point of divergence amongst divers when it comes to the question of the future of cave diving. Some feel that new blood should be injected into the sport, that every encouragement should be given to any caver showing interest in diving – provided that he follows the sensibly searching and rigorous training procedures laid down by the Cave Diving Group. Others believe that cave divers who are going to make the grade should be independent enough not to need any encouragement from outside. Boon belongs firmly to the latter school of thought, and believes he is morally bound to actively *discourage* people from diving. To him, the phrase 'new blood' – if it consists of young cave divers eager to please their mentors – seems all too apt.

Whether would-be divers are to be encouraged or not, the men involved in this particularly hazardous aspect of caving agree on one thing: if a caver *does* insist on taking up diving, then he should be properly trained. The idea of someone inexperienced strapping on an aqualung and jumping into the nearest sump does not bear thinking about – yet it happened in 1970, with appalling results.

Alan Erith was in his early twenties, had six years caving experience behind him, and was terribly eager to start cave diving. Whether proper training was ever suggested to him is not known, but it is know than on Friday 2 October he bought a full diving outfit, and the following day made his first dive in Keld Head Rising, the sump dived earlier that year for nearly 1,200 feet by Mike Wooding. Perhaps it was news of this epic dive which played a major part in his choice of location, and possibly even in the evident enthusiasm he displayed.

By way of initial training, Erith jumped into the pool outside Keld Head, swam round a little, then dived for the very first time. He followed the line left by Wooding for about 30 feet, lost it, then returned very shaken by his experience to the surface. There he calmed down by having a cigarette with his friends, then dived again. Those on the surface felt the line twitching for about 20 minutes. After 30 minutes, the Cave Rescue Organization was called out. For Alan Erith, the callout was far, far too late.

His body was never found. Wooding, the first diver on the scene after the callout, immediately checked the first airspace. There he found the line very slack and badly tangled. As he prepared to dive again, he realized that the line was in fact completely slack and, pulling it in, found that the end appeared to have been freshly cut.

Even highly-trained divers can have problems with loose lines, and the most likely explanation of the tragedy is that Erith became entangled in the line, managed to cut himself free but in the process lost his mouthpiece.

October of 1970 was a black month for cavers in Britain. It was as if they were being taught that cave water, with all its fascination and mystery, was still very much in charge of the proceedings. Only two weeks after Erith's disappearance came the second blow.

Stephen Sedgwick, an 18-year-old university student, was making his way out of Porth-yr-Ogof, a popular river cave in South

Wales, and only the second he had ever been down. Two of the party of four, including the leader, decided to exit from the cave by swimming out through the deep but open resurgence. Sedgwick followed them. Suddenly, he cried out, then disappeared underwater. Jim Delderfield, the leader of the party, could see the electric light of the drowning man 15 feet underwater, and made repeated attempts to dive down to him. Ironically, the buoyancy of his wetsuit – normally a blessing when swimming underground – prevented him from diving deeply enough. Experienced divers were at Porth-yr-Ogof within 50 minutes of the accident, but by then the cave had well and truly claimed its fourth victim by drowning, and not the last . . .

The death of Stephen Sedgwick was not that of a cave diver, but of someone who fell prey whilst pursuing the *technically* straightforward practice of underground swimming. It is included in this chapter because it proved to be the link in three incidents which, chiefly because of their close timing, shook cavers and cave divers to the core.

Paul Heinz Esser was no newcomer to either caving or diving, and displayed such a natural ability underwater that some believed, given time, he would reach the standard set by John Parker, probably the most successful cave diver to emerge in recent years.

On Saturday 13 February 1971, a few weeks after he had been elected a diving member of the CDG's Somerset Section, Esser dived in Porth-yr-Ogof. His objective was to enter the cave by a low entrance until he reached the Rawlbolt Air Bell, then make his way out via the Top Entrance, reeling in a loose line as he went. For some reason, the vital line he should have followed to the Top Entrance had come adrift by that entrance and floated downstream, back into the cave. Divers do not secure their lines with inadequate granny knots – particularly Parker, who had laid the line originally – and there is a strong case for believing that some casual visitor to the Top Entrance may have untied it out of idle curiosity.

Whatever the cause, the result came with horrible inevitability. Esser followed the trailing line, believing it would lead him to the surface, but now the picture was different – it led him to a completely flooded new section with no air spaces. It was here that his air ran out.

After one of the largest underwater searches ever attempted, Paul Esser's body was found the following day, his hand still firmly gripping the line which he had believed would lead him to safety.

No cave diver follows his chosen pastime for long without at least one near shave, and it is the hallmark of a competent diver that he can deal with minor mishap after minor mishap as they arise. He knows full well that, unchecked, their total can spell the end for him.

Cool-headedness brought Adrian Wilkins out of a decidedly nightmarish situation in Ogof Afon Hepste, South Wales. When 250 feet into sump three his apparatus developed a high-pressure leak. Repair underwater was out of the question, so he turned back for base, operating the tap by hand so that he could take a breath every half minute. After about 150 feet of the return journey, his helmet came off. Patiently, with one hand controlling the air tap, he found and replaced it. Making his way up through a slit, his face-mask was torn off. He groped around, found it, cleared it of water and carried on. No doubt to relieve its owner's boredom, the demand valve chose this time to malfunction. With his air supply down to a few breaths, Wilkins surfaced at the base, no doubt blessing the CDG policy of diving with a 100 per cent air reserve.

Cave diving is popular in most of the caving regions of the world, but it is interesting to note a quite basic difference in attitude between British and American divers. In the USA, diving and not caving is the basis for cave diving – the already experienced diver must be taught how to cave. The British philosophy is that a sump is a cave passage which happens to be filled with water, that the main object of cave diving is to find more cave, so here it is the experienced caver who must be taught how to dive.

Which is the correct approach is obviously open to a great deal of trans-Atlantic debate, but it is interesting – and horrifying – to note that in the ten years from 1960 to 1970, Florida alone produced 84 cave diving deaths. Admittedly, many of these were reportedly caused by non-caving divers finding the sea too rough for the day, and deciding to investigate one of the enticing warm and clear springs of that state.

How hazardous is cave diving compared with ordinary caving? Boon was blunt: 'Bloody hazardous, by a factor of many.'

Regardless of the risks, cave diving will – and must – continue. Jochen Hasenmayer's record dive of 2,400 feet into the de Balme caves of SE France will be passed, and again, for this special breed of caver holds in the steel tanks strapped to his side the key to hundreds of miles, perhaps thousands, of otherwise unreachable cave passage.

Chapter 7

Underground Hazards . . . and Rescue

The most dangerous part is the drive to the cave. It sounds clever, and to a large extent is quite true. Road accidents to cavers have in all probability kept pace with accidents underground. But the statement is a sham one in its implication of relative safety inside the cave; traffic hazard is something we have all got to live with, whether caver, mountaineer or tiddley-winks devotee.

To take up caving is to deliberately select a leisure activity with a higher degree of risk than many other sports, so the question 'Why?' is not an unfair one. Some of the varied reasons cavers have for taking up their sport were given earlier, but the presence of real danger is undoubtedly one of the strongest.

Perhaps it seems odd that anyone in this welfare-stated, well-policed world should go out of his way to risk his neck, but I am convinced that it is this very mollycoddling which is responsible. When, tens of thousands of years hence – unless the test-tube geneticists speed things up – human beings have evolved into near-limbless and non-muscular brains in basket chairs, there will be no need for such activities. We will get complete fulfilment from our man-made environment. But present-day man, despite his assertions otherwise, is removed by just a sliver of time from those hirsute ancestors whose one aim in life was to keep on living. Our bodies are still being made to pretty much the same design, with muscles for transport, attacking and defending, and a host of glands ready to pump adrenalin and God-knows what else into our bloodstream for do or die survival situations. Our brains have certainly acquired a deeper subtlety and intellectual ability, but still retain the mental machinery for coping with day-to-day threats to life and limb.

Unless he deliberately puts himself in a risk situation, twentieth-century man in the western world has very few opportunities for using this survival equipment. While the physical components may atrophy without bothering the owner, this is not the case with the brain. There is plenty of evidence in society to demonstrate what happens when our survival-orientated minds meet with a bland and risk-free environment – delinquency, vandalism, pointlessly violent crime and unwanted physical aggression come to mind.

Thus, if the civilized world does not make sufficient demands of you, adventure sports will. And of these, caving is both cheap and fairly convenient geographically.

It is difficult to place caving into any scale of risk, but it hovers perhaps somewhere between ski-ing and mountaineering. A would-be suicide looking for somewhere to do the deed would find all the requirements in any decent pothole, but cavers do not come into this category and have evolved equipment and techniques which ensure that they keep out of trouble most of the time.

What are the risks in caving? In Britain and on the European Continent – and possibly Canada as well – the biggest single danger is water.

Underground water – a seductive black liquid which, like a loose-tongued mistress, can provide boundless pleasure or the gravest peril. Without water, a cave seems – and is technically termed – dead. The passages, shafts, crevices and enormous halls which were patiently carved from the rock by countless million gallons of water now seem as lifeless as a fireplace with no fire. But take a trip through a living cave and you experience the fullest pleasures of the sport, a witness of nature going about her gigantic task of making the earth. The scale of cave water at work varies from the steadiest drip-drip-drip of stalactite stretching for stalagmite, to the thunderous clamour of a flooding river hellbent for the bottom of the cave . . . and beyond.

The amount of water entering a cave can vary a great deal in a short space of time; two or three days of rain can leave high moorland well soaked, ready to precipitate a flood into the caverns below if even a light rainfall adds to the general water content of the soil. Equally, the end of a drought can be heralded by underground flooding, as the parched and stone hard earth lets the rain follow a rapid downward course. A careful check on the local weather forecast is vital for any caving party intending to descend

a flood-prone system, but although this safeguards against predict-
able bad weather there is always the danger of a sudden and quite
unexpected rainstorm hitting the area. Nor does the rain have to
fall near the mouth of the cave for it to present a hazard; those
caves which are fed by streams or rivers whose headwaters are
some way away can be subject to flooding if heavy rain falls
at the source. Little Neath River Cave in Wales is such a system,
the entrance to which is a cleft in the bed of a river with the water
at normal level lapping only inches below the lip. It always gives
me a shudder when negotiating the first hundred feet of tight and
jagged passage to imagine the sensation of trying to hurry through
this part – rather like those dreams in which one tries to flee
through a river of treacle – with the rising river pouring into the
entrance like a giant, emptying bath. Nigel Clarke, a caving friend
who did have to exit from Little Neath in rising flood conditions,
reports that the level shoots up with alarming speed, and while
he exited with a little airspace left his companion – understandably
very close on his heels – had to be pulled up to the daylight through
a solid vortex of water.

When the unexpected rainstorm does arrive, the leader of a
caving party has a choice of two courses of action: to sit out the
flood in some high and dry passage, or to make a dash for the entrance. Neither course is without risk. The lower reaches of many
caves and potholes, and sometimes entire systems, can flood completely; if the decision is to sit out the flood, the leader must be
sure that his chosen site really is high enough to escape the water.
As he can never be sure of how great the flood will be, what may in
nine cases out of ten be a safe refuge could flood to the roof on the
tenth. He must also take into account the grim fact that the flood
may persist for many days, although as often as not they subside
as quickly as they rise.

The alternative of making a dash for the exit is even more full
of doubt. The team may get as far as just a few hundred feet from
the exit to find that last stretch completely submerged, and the way
back down to a safe refuge already cut off. In a pothole, there is
the added danger of having to climb ladder pitches against a possibly overwhelming force of water.

I learned a hard lesson once when coming out of Penyghent
Pot with the water rising to a moderate flood level. Ladder climbs

which were normally quite easy, despite a thorough dousing, were transformed into desperate struggles. The very bottom pitch, aptly named Niagara, bore a column of water as solid as though the end of a completely full bath tub had been suddenly chopped off; once inside the cataract, clutching desperately and blindly at each thin rung, there was no way of contacting the others who were double-lifelining me up, so overwhelming was the noise. Some cavers trim off their helmet peak to give the headpiece a nattier look, but I was glad of the peak on mine for it created an area of foam in front of my face, rather than solid water, in which I could breathe through pursed lips. The pitch was one of the shortest in the pot, but made every one of us fight for our lives up every single inch. Add to this the gradual numbing effect of the cold water on the arms and the hands grasping each metal rung, and you will have some appreciation of why it is often better to sit and wait.

Even at normal level, cave water can be treacherous. Usually the only way to make progress through a really wet cave is by plough-ing through the water; often the stream is only ankle deep; some-times it plummets to very considerable depths. Where the appar-ently easy stream passage contains flooded rifts – which in the light of the caver's lamp do not show up – the caver must proceed par-ticularly carefully, testing every step. The stream passage in Ogof Ffynnon Ddu, South Wales, contains many such traps, and steel poles have been laid across them; the newcomer to OFD is often surprised to see the leader plodding heavy footed through the water then suddenly going into his tightrope balancing act.

Where the deep parts are too extensive to traverse round or stick poles across, a team has either to swim or take to the boats. The only practical boat underground is the rubber dinghy, but its thin skin is all too easily punctured on rocky spikes and razor sharp edges. If fully wetsuited, a spill into the water might not seem too serious, but in truth swimming in full caving gear – probably with a haversack round your shoulders as well – is difficult. Under-estimating the hazards of subterranean swimming led to the death of Stephen Sedgwick in Porth-yr-Ogof.

As though to emphasize the point, the news reached me as this chapter was being written of another cave drowning tragedy, this time in Scotland. A caving team from Bristol was exploring Fire Hose Cave and making a diving attempt on the terminal sump.

7. The wetsuit has greatly improved the caver's resistance to cold in such decidedly wet situations as this climb in Bull Pot of the Witches in the northern Pennines

8. The breathtaking grandeur of this, the final shaft in Juniper Gulf, one of Yorkshire's classic potholes, helps to take the caver's mind off the sheer slog of a 175-foot ladder climb

Whilst the dive was in progress, Peter Clements started out of the cave with another companion; neither wore a wetsuit, and both were feeling the cold waiting by the sump. At the entrance, Peter fell into a pool eight feet deep and about fifteen feet wide. In his exhausted state he panicked and could not regain the surface. The other caver jumped in but was unable to help and only just managed to get out of the pool himself.

Such are the dangers of cave water in the straightforward and immediate sense, but it is when added to exhaustion and hunger that it becomes even more ominous, for this is the mixture which will lead to the worst caving danger of all: exposure.

Exposure, which the medical man will refer to as hypothermia, is an insidious killer which lies in wait in every wet and cold cave. It is not an ailment which suddenly manifests itself out of the blue, but rather begins to work on any caver, however fit, the moment he crawls in through the entrance. Its killing action is simply a lowering of the vital core temperature of the body (that of the trunk and brain); while the limbs and skin can drop many degrees without undue damage, the core temperature cannot fall by more than a very few degrees before unconsciousness and death follow in rapid succession.

The constant cold and wetness which a caver faces on an average caving trip, plus his high energy burning rate, demand constant caution. Not only must the leader of a party keep an eye open for the first symptoms, such as slurred speech or poor co-ordination, in others, but each caver must hold the individual responsibility of knowing when to call a return to the surface. This is particularly difficult for an inexperienced caver. Firstly, he has a feeling that he may be letting the side down (not realizing that he certainly would be doing so if he left it too late and had to be rescued), and secondly, it is often difficult to differentiate between simple tiredness and the onset of exposure.

Once it has a firm grip, the full effects of exposure strike with appalling speed. A caver in a party I was leading down Penyghent Pot, whilst no beginner in the sport, was tackling his first really serious pot. About half-way down I had a quiet word with him, and he assured me he was feeling as fit as ever. But on the very next pitch his strength failed him, and I called an immediate return to the surface. In the few hours it took us to exit, I watched him declining at an increasing rate; we very nearly reached the point where I would have had no other choice but to send for help from the outside.

On another trip, this time into Cwm Dwr Cave, South Wales, my companion badly wrenched his knee crawling through a boulder choke. I pulled him out of the constricted choke just before his knee locked solid, made him as comfortable as possible, then went for help, leaving him in some pain but otherwise perfectly fit. Fortunately I met a party making their way into the cave and two of them were able to go out for help while I went back into the cave. In the half-hour of my absence, shock and the onset of exposure had reduced my injured companion to a shivering and glassy-eyed state.

Both these men escaped with their lives, but exposure has claimed more lives in caving than anything else. Harold Sargeant, a 30-year-old Sgt-Major who had just passed his medical for another term in the Army, found himself weakening on the return journey in Yorkshire's Grange Rigg Pothole, a difficult series of ladder pitches. Asked if he could manage, Sargeant said he could make his way out, but in truth his life was draining away. He died from exposure not far from the entrance. 20-year-old John Williams also followed the same deadly pattern of exhaustion and hunger, having eaten very little the previous day, on his way out of Penyghent Pot.

Even more rapid was the fate of a girl caver in Longwood Cave, Mendip. She had been too tired to climb a short drop only minutes from the entrance, and stayed there while other cavers went out for help. Close to the entrance a cave rescue team was already changing into their caving gear, ready to set out on a training trip. They were back down with the girl in a very short time, something of the order of fifteen minutes from the time of her first difficulties – but time enough for exposure to stake its claim on another life.

It is commonly supposed that progress through a cave must be accompanied by coughing and retching as pockets of noxious gases are passed. In fact, the air in most caves is remarkably pure and sweet, kept circulating by water movement and natural convection. Foul air is very rarely encountered, although there are a few well-known spots where air movement is non-existent – such as between sumps – and where there can be a build-up of carbon dioxide if the caver stays there too long. Some small digs are subject to this build-up, too, and a common practice is to leave a candle burning as you work: when it goes out, so do you.

Rock falls are also much less common than non-cavers believe. A water-worn passage through solid limestone is, let's face it, about as likely to collapse of its own volition as a concrete drain pipe. The real danger of falling rock is at the bottom of pitches when rope, ladder or climber movement may dislodge some delicately poised boulder (hence the golden rule of never standing at the bottom of a pitch gawping up as the next man comes down, however impressive a sight).

Boulder chokes or ruckles present considerable risks, too – those chambers which have become partly or completely blocked by fallen boulders from the roof, too large to sustain perfect support. Some well-worn ones which have witnessed the passage of thousands of cavers are pretty stable affairs, while others are negotiated with cat-like movements and ne'er a whisper. Partial collapses of chokes have caused many a nasty turn, and more than a few bruises, but relatively few bad accidents. One exception is the entrance choke to Eastwater Cavern on Mendip, a limestone jumble of no more stability than a playing-card house. Here, a caver received fatal injuries from a shifting boulder.

There is only one possible approach a caver can take in a dangerous boulder choke: a highly philosophical one.

One extra caving hazard shared by America and South Africa is the disease histoplasmosis, an infection of the lungs rather similar to tuberculosis and long classified as the mysterious 'cave disease'. Then it was tracked down to the spores of certain fungi which thrive on the droppings of bats and other creatures. In Britain, our caves are pleasantly free from such horrors, but one caver did contract an infection in the Mendip cave called Stoke Lane Slocker which could have had very serious results. Fortunately, the caver in question was also a pathologist and was able to recognize it as Weil's Disease, an infectious jaundice transmitted in the urine of rats. Prompt treatment saved him, and cavers were made aware of this additional hazard in cave water contaminated by farms and other habitations.

Often it is possible for a team to evacuate one of its own

members suffering from a minor accident, but a more serious injury means a full-scale cave rescue callout.

With its fairly compact caving regions, Britain boasts one of the most efficient rescue systems in the world. Such systems exist in other main caving countries, too, but there the greater distance to be travelled to the cave can cause more serious delays.

A cave rescue callout in Britain is always made through the police, at their specific request, and they have the final say in most matters although obviously the experts are left to get on with the job. It is the police who, in extreme cases, will call in extra surface assistance from the fire brigade or the Army. Such help is needed when a stream has to be diverted from the mouth of the cave by damming and pumping; such diversions will only clear the flooded cave if the water is pumped into some course which does not lead back into the same system.

As soon as the police are alerted, they inform the caver responsible for calling out however many rescuers are considered necessary. A snowball and standby system is operated, so massive additional help can be obtained from other regions if required. Those who volunteer to serve on a rescue team have to be prepared to leave their jobs, beds, meals and pubs at a moment's notice. A really big rescue, stretching over several days, can entail a considerable loss of earnings for some, but they never grumble. As cavers themselves they fully appreciate the need for a rapid evacuation of the victim, for time is the greatest enemy of a man injured deep inside a cave, time in which exposure and shock can finish off the work of the initial injuries.

Cave rescue techniques vary from operation to operation. Where there is crawling or standing room, the victim is strapped into a modified Neil Robertson stretcher, a bamboo and canvas affair originally designed for work on ships. Where space is constricted, or there are many bends to be negotiated, he is placed in a canvas drag sheet. Unlike mountain rescue, in which it is usually possible to get a fairly large team manhandling a stretcher, constricted cave passages often dictate that only two men can perform the enormously tiring task of manoeuvring it – one at the front, one at the back. Pitches are tackled in advance by a separate tackling party, so no time is lost in hoisting the stretcher up by a pulley system. In high and narrow rifts, or partly water-filled crawls, rescuers act as a human bridge for the victim to be passed over, the end man climbing up and over the stretcher as it passes over him to take his place at the head of the bridge. While deep water can be negotiated with the stretcher on one or more dinghies, sumps present a much more formidable problem. Fortunately, the Cave Rescue Organization now has access to special equipment which enables an unconscious man to be taken through a sump – and still breathe.

A few hours' stretcher work on top of the effort needed to get into and out of the cave is about the limit for a rescue team, so fresh men have to be sent in at carefully timed intervals. In such a complex operation, the surface rescue organizer must be kept right up to date with events underground, so good communications are vital. Radio is out of the question, except in shafts, so one of the first tasks in a rescue operation of any size is to lay a telephone cable.

When the first rescue team reached my injured companion in Cwm Dwr, mentioned earlier in this chapter, I started to make my way out of the cave. As I crawled through the long and low entrance passage, I was aware of a peculiar noise ahead of me, rising quickly in volume. I rounded a bend, and what I saw made me blink: coming towards me at an incredible speed was a caver – on his back! He was the telephone man, laying his precious cable out of harm's way in cracks and crevices, and moving at breakneck speed all the same.

Progress in the development of caving techniques and such equipment as the wetsuit has led in recent years to the exploration of caves which would once have been classed as impossible. With such an extension of the standard of difficulty comes a massive problem for the Cave Rescue Organization, for some of the ultra-difficult systems present rescue problems on a scale never tackled before.

Towards the end of 1970, the CRO prepared a 'Black Book' of 56 caves in the north of England alone from which a seriously injured caver could not be extracted by conventional rescue techniques. Cavers being cavers, this has not led to any reduction in the number of visits to these systems, but it has resulted in the CRO taking a completely fresh look at their techniques. To meet the challenge of the first official rescue from one of these blacklisted caves, contingency plans are being made to transport National Coal Board drilling derricks to chosen sites above the cave. So constricted are these systems that it is probably quicker to drill an escape shaft down to a chamber whose precise position is known than to attempt to blast away the constrictions. Another line of approach which must now be seriously considered is the underground hospitalization of the victim until a shaft is drilled, the passages cleared, or he is well enough to cope with the gruelling punishment of a normal exit. Such an approach would involve the erection over the victim of a tent which would be kept heated, and constant medical attention over perhaps several days.

Cave rescues make big news, and attract far more adverse publicity than the sport deserves. So the next time you see those familiar glaring headlines, remember what really lies behind the story: not a despairing cry to the outside world for help, but cavers helping cavers.

Chapter 8

The Blackest Hours

The entrance to Peak Cavern, in the heart of Derbyshire's caving country, is magnificent by any standards. The enormous gash cutting into the heart of a hillside, surmounted by Peveril Castle, is the limestone prelude to a few hundred yards of large tourist cave then some miles of wild cave, difficult enough to captivate any sporting caver. In 1957, a low and muddy bedding plane, Pickering's Passage, was discovered leading north from the main cave; two years later, local cavers pushed even harder, and found another 100 yards or so of passage, some muddy chambers and a T junction. On 8 March 1959 one of the passages leading from the junction was pushed down to a pool and then up again into a chamber. Leading from the chamber was an almost vertical shaft heading downwards from a fireplace-shaped hole in a slope. The spot was earmarked as a likely continuation, and a possible link with the large series of passages which are believed to lie beyond this limit of Peak Cavern.

On Sunday 22 March two teams went into the cave, one to the known parts of the system and the other to take a closer look at that shaft at the limit of exploration. The leader of the latter party, John Randles, included in his team a 20-year-old philosophy student from Oxford who had asked to be allowed to go along. Randles had not met Neil Moss before, but was quite satisfied as to his competence.

Randles' team worked their way into the new section and lowered 75 feet of electron ladder down the virgin shaft. With his medium height and slightly slim build, Moss was quite happy to be the first down the shaft, and no doubt was savouring the prospect of new discoveries – perhaps *the* discovery, that dream of all cavers. Although the team had a handline for the slope leading up to the shaft entrance, no lifeline was used on Moss: the shaft was so tight that it was reasonably supposed that a fall would be nearly impossible, and anyway a rope would tend to jam.

Moss quickly found that the descent of the shaft was a far from straightforward affair. Some ten feet down he met a corkscrewing constriction and wormed his way through this, feet first, to another drop. At a depth of 35 feet, his boots came across an obstruction of boulders and he began to kick at this in an attempt to clear the way down. Not only was Moss unable to move the obstruction, but his efforts had also jammed the ladder. The cavers at the top of the shaft tried tugging the ladder free from the boulders, while Moss managed to support himself by some other means, but without success. Having done more than his fair share, Moss was told to climb back up the shaft so that someone else could have a go.

After some while, those at the top of the shaft grew puzzled at the non-appearance of the young student, and at the scratchings and scrapings which told of some difficulty. Moss reported the situation: so narrow was the final stretch of shaft down which he had slid that he was unable to bend his knee to get his foot into the next ladder rung. Nor could he raise his legs to get some kind of purchase on the walls.

Things were awkward, but surely not insurmountable. Moss was not held in anything resembling a vice-like grip, but simply unable to raise himself those few inches necessary to get his foot on the next ladder rung. As he was unable to help himself for the moment, obviously assistance had to come from above. One of the team climbed down into the shaft as far as the corkscrew and heaved on the ladder with all his strength, which was considerable. Moss was, it seems, moved a short distance – then the ladder went tight and quite immovable, its lower end still jammed solid by the boulders.

Attempts to lift Moss on the ladder were obviously doomed to failure, so Randles quickly arranged for the rope handline to be dropped down to the trapped man. Those at the shaft head heard his violent struggles to get the rope under his arms, but there simply was not enough room. The rope was pulled back up, a loop formed in it, and by some extraordinary effort he managed to get this round his body. The team began to pull, and it seemed as if the Neil Moss story was to be just a few interesting lines in the chapter of near-misses, a chapter contributed to by just about every active caver at some time or another.

Then, without warning, the rope snapped. Neil Moss fell back those hard-gained few feet, down to square one, and began to die.

The rope which was used on that first attempt was manilla, a natural fibre. If the Neil Moss episode had taken place today, he would in all likelihood have been hoisted to safety with just a few heavily bruised ribs to show for his troubles – for today, that rope would have been fashioned from nylon, or some other strong man-made fibre.

Back in his own private hell, Moss faced an immediate problem even greater than exposure or lack of food and drink. Ventilation in the shaft was almost non-existent, and gradually his own exhalations were poisoning the air. In addition his acetylene lamp, the flame now extinguished, was pumping out the anaesthetizing gas to make matters worse.

Another couple of ropes were hurriedly brought into the cave, but cavers attempting to go deeper than the corkscrew had to be pulled back up as they approached the verge of blacking out in the foul air. Neil Moss passed into unconsciousness – his last intelligent conversation with those above had been about the accommodation he had arranged at a farmhouse – and this was both the worst and the best thing that could have happened. Unconscious, his breathing dropped to a rapid panting at a lower level than if he had been conscious, thus whatever oxygen was present was used at a lower rate. Unconscious, he was spared the living nightmare of being fully awake to the fact of his entombment during the prolonged efforts carried out to save his life. Yes, Moss was well out of things in his state of oblivion . . . but no longer was he capable of passing a rope loop round his body. That operation had to be done from immediately above him, and for a long time no caver could pierce the barrier presented by the foul air in the tight pit.

At 2 a.m. on Monday morning, a caver by the name of Ron Peters was awoken by the police who told him of the rescue, and the desperate need for small, highly experienced cavers. Without hesitation Peters volunteered his help, and was in the cave at the head of the shaft two and a half hours later. The other rescuers looked at this new arrival with renewed hope, for he stood barely five feet in height and weighed only seven stone.

Stripping off his outer clothing, Peters quickly worked his way down the shaft and through the tight corkscrew. Then that final section of shaft. Between his teeth he clenched a rubber tube, kept supplied with oxygen from a bottle at the shaft head. Peters' boots made contact with something which yielded slightly, and by straining his head to one side he saw below him the helmet of the trapped man.

How a man of even Ron Peters' slight build could have managed it in that constricted fissure is beyond imagination, but by some supreme effort he picked up the broken rope around Moss's body and tied a fresh one to it. On the verge of blacking-out, Peters fought his way back up the shaft – there was some fault with the oxygen supply, and only by sheer determination did he reach the top, where he collapsed. But Peters had done it, had beaten that pool of carbon dioxide in which by some miracle Moss still breathed. The cavers took the strain on the fresh rope, and by inches Moss started his journey to safety. Then, irrevocably, he jammed, having been hoisted perhaps 18 inches.

Only one other caver managed to reach Moss again, a slim and completely determined 18-year-old girl, June Bailey. Like Peters, she went down feet first and got near enough to see that Moss had one arm jammed under a rock projection – however hard the rope was pulled, that arm would act as a wedge and prevent any further upward movement. June climbed back up the shaft and insisted on being lowered head first so that she could move Moss's arm and fix another rope around him. John Randles was willing to try almost anything to save the life of his fellow caver, but not at the expense of another life. And there can be little doubt that a head-first attempt down that shaft would have been a one-way trip.

In the outside world the biggest cave rescue operation ever was being built up. Private companies maintained a constant supply of oxygen cylinders so that the life-giving gas could be piped down to Moss; volunteers came from the Mines Rescue Organization and the Navy as well as from other caving regions; those who lived in the nearby town of Castleton opened their doors to exhausted rescuers; the wvs were there, of course, providing the hot food and drinks needed to keep the operation going; the police assisted with communications and struggled with the ever-increasing flow of traffic on the roads to the cave. The outside of Peak Cavern was the thick end of a wedge of people in ceaseless activity, all their efforts devoted to one very inactive man at the tip. Given time, men can quarry away a whole mountain – but at Peak Cavern precious time was ticking away with every extra buildup of carbon dioxide.

Time ran out. Some 32 hours after he entered that fireplace-shaped hole deep under Derbyshire, Neil Moss died. Attempts to recover the body failed, and the entrance to that fateful tube was sealed with concrete.

Many bitter lessons were learned during those few days in the shadow of Peveril Castle. Strong criticisms were levelled at the rescue organizers in some quarters, and there were accusations of the rescue attempt being stunned into a mindless muddle by the sheer drama of the situation. Why was more attention not paid to sucking out the carbon dioxide from the shaft when it was well known that oxygen pumped down would simply rise up again? Why was no immediate attempt made to tunnel to the trapped man when this would not have interfered with the efforts being made in the shaft itself? Valid questions indeed, but at least the lessons had been learned, the hard way.

Getting stuck or washed away by floods are the hazards one would expect to be most common in caving, but it is the cold and the wet – or more particularly a combination of the two – which cavers are most wary of. Loose stones can usually be spotted and avoided; lifelines can be used on pitches or high traverses; caving can be called off for the day if bad weather threatens. Exposure – the relentless debilitation of the body through a combination of cold and wet conditions leading to a cooling of the body temperature; exposure is something different. Knowing whether one is simply cold and tired but safe, or cold and tired and declining, takes a great deal of experience, and the most disturbing fact of all is that often a victim of exposure is incapable of judging his true state of health. It is rather like those winter mornings when one stumbles out of bed and finds it hard to tell whether the muzziness is due to the beginnings of a cold or just the morning-after feeling.

John Williams left the decision too late on a trip down the super-severe Penyghent Pot. One of the toughest systems in **Britain**, Penyghent is not an undertaking to be approached lightly, and yet for some inexplicable reason Williams set out on the descent without having eaten. The team had just started the ascent when Williams began to feel dizzy and overtired. The feeling of depth and isolation from the world above is particularly strong in this pothole, and the knowledge that he had all those pitches to fight up plus that final, relentless thousand-foot crawl must have shocked him even more. Sometimes the victim of exposure can be talked out of the last stretch of passage, but no one could pretend to Williams that the climb which lay before him was easy.

The team was able to haul him up the smaller pitches, and by the time he reached the foot of the main pitch he seemed to have recovered a little. But the free-hanging 70-footer, although rigged clear of the water, took too great a toll. When he crawled on to the tiny platform which splits the 70 foot pitch from the 60-footer above, Williams was completely exhausted, all his reserves spent.

That platform dividing the two largest pitches in Penyghent is the coldest place I have ever come across in any cave or pothole. The water thunders down the top pitch on to the small area of level rock, no larger than an extended dining-room table, and it is impossible to escape the numbing spray. Just as bad is the fierce draught generated by the waterfall and this, combined with the wetness, makes the platform a decidedly unhealthy place to linger long on.

Williams made a feeble attempt to climb the 60, then collapsed on the platform. The leader hugged him in a last despairing attempt to impart some body heat, but, amidst the clamour of the water, the cold sucked out the last spark of life.

Only a month after the death of Williams, Penyghent Pot seemed set to claim yet another life. John Frankland, 30, from the Yorkshire town of Keighley, descended the eighth pitch without a lifeline. He fell, some 20 feet, and much of the force of the fall was absorbed by the shallow pool at the bottom of the pitch. But there were minor fractures, and prospects for getting Frankland out of the pot alive looked bleak. Fortunately, he was a caver of remarkable stamina and courage, and did emerge alive – but only after a 26-hour rescue.

Mine exploration is closely allied to caving, but has its own particular hazards, especially in the way of gas deposits (extremely rare in caves) and unstable rock. Old mines have also taken a much greater toll of life amongst non-cavers – the schoolboy clutching a bicycle lamp variety – than have caves, simply because of their more obvious entrances, relatively unconstricted passages, and their often close proximity to centres of population.

Of all our old mines, those riddling the sandstone of Alderley Edge in Cheshire have claimed most lives, something like a dozen at the time of writing.

Thousands of youngsters must have groped their way through the Alderley Edge mines, and possibly hundreds have had to fumble their way out blindly as their matches, candles or torches gave out. Many have got away with it, making their escape to the daylight to brag to their friends about their adventure. Twelve have not.

Alfred Hadfield and his friend George Etchells, both 24 years old, were rather more than youngsters. And yet they took the appalling chance of exploring deep into Alderley Edge with just a few candles one Whitsuntide afternoon. Nine days later – why so long it is difficult to ascertain – their absence was reported by their families, and a thorough search was made of the mines. But the delay had been too long ...

The search parties could find nothing. Then, the following September, a team was exploring a part of the mines known as the Extension. The leader climbed a 60-foot shaft, and there found the two almost unrecognizable bodies of the missing men. In the perpetual darkness of the mines, Hadfield had scrawled a note on the back of an envelope: 'Dear Mother. We are lost in the copper mines. I think we are done. We are . . .'. The letter was never finished, and possibly it is as well that the parents were spared the thoughts of the two at that time, for they were dying in the manner of man's worst nightmares.

So the list of underground fatalities is added to, year by year, but the most recent figures do not show any startling increase in the percentage of accidents. It is simply that more and more people are caving, and that new developments in safety, such as the now widespread use of the wetsuit, are balanced by exploration of more and more difficult caves. Such a cave is Mossdale Caverns.

When I was co-editor of the caving magazine *Speleologist*, I received a personal letter from David Adamson, a member of the University of Leeds Speleological Association. Dave was an expert on the inner reaches of the six-mile long Mossdale, and in his letter he asked me to emphasize through the magazine the extreme flood dangers of the cave. He was well aware of these, and wanted to make sure that the message was widely broadcast.

Mossdale is a great, ultra-demanding system, compared by some as a British caving equivalent of the Eiger's North Face. Certainly, the question of objective hazard is just as great: stone-fall on the Face, flooding down Mossdale. Dave, and his companions, knew of this risk, but flood danger has never stopped any exploration for very long. It means simply that one must be even more choosy about weather conditions on the day of a descent.

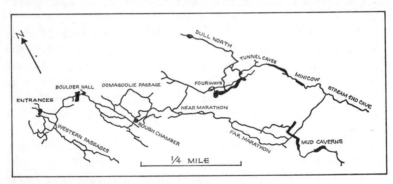

The complexities of Mossdale Caverns

At 2.30 p.m. on Saturday 24 June 1967, a group of cavers went down Mossdale Caverns. When they reached a place known as Rough Chamber, six – the most experienced – continued into the heart of the system. The others made their way out of the cave. The group of six, led by Dave Adamson, made their way through the low wet Near Marathon and Far Marathon series to survey at the farthest known points of the cave.

The party which left early included Miss Morag Forbes, Dave's fiancée, and when a storm broke later that day she ran to the entrance. Mossdale Beck was in full spate, the entrance to the cave awash in the fierce torrent. She immediately alerted the rescue teams, and that night and the following day saw the most furious efforts made to divert the water from the cave. Bulldozers and earth diggers were used to dig a huge trench, and twenty fire-pumps worked non-stop at reducing the level. An attempt to reach the men was made from the surface as early as 2 am on Sunday, but this was thwarted by the sheer volume of water.

Later on Sunday, the first teams were able to make their way through the Marathon Series where, eighteen hours after the rescue attempts had started, the full terrible truth was discovered. All six men had died, unable to make their way against the water to those parts of the cave where they would have been safe.

With the unleashing of that thunderstorm, the blackest chapter in our caving history was written. The bodies of the six were beyond recovery, and the entrance to their cave was sealed. With the full consent and blessing of the parents concerned, nine cavers paid another vist to Mossdale in a dry spell in 1971, but this trip was to draw no attention from any outsider. The team recovered the remains of their six friends and buried them in a high-level cavern in the nethermost regions of Mossdale.

One of the world's most bizarre caving mishaps occurred in a small Austrian cave, the Frauenmauerhöhle, at the end of the last century. So bizarre, in fact, that the incident is almost like an episode from black comedy, but the outcome was far from humorous.

Three students decided to explore the cave, a well-known and technically easy one, and were proceeding quite happily until, half-way between the two entrances, their torch went out. They had a supply of matches, though, and were fairly confident of being able to complete the through trip without difficulty. As the light cast by a match is at best rather limited, especially when three men have to progress by its glow, they decided to follow one of the walls. As each match went out, they were able to grope their way along the chosen wall for a few feet before lighting another.

How long they carried on in this fashion it is difficult to say, but one can assume that as the supply of matches ran down, so they relied on following the wall for longer and longer periods in the dark. The last match was struck, the students dashing forward as fast as possible in its feeble glow, then it, too, spluttered out. The utter blackness of the cave descended in an instant, and the luckless three had to rely on the knowledge that if they followed their wall, even in the dark, sooner or later they would come to one of the entrances. After all, they must have kept telling each other, the cave wasn't all *that* large.

They never did emerge from that oh-so-small cave – at least, not alive. Their bodies were found, close to the wall in which they had pinned all their faith, and only a short distance from where they had struck their first match. A brief inspection revealed the key to the tragedy, and in the bright lights of torches it must, at first, have been difficult to accept: the 'wall' which they had faithfully followed belonged to an enormous rock pillar, 150 feet in diameter. And round that pillar, at regular intervals, a trail of spent matches.

Chapter 9

On the Brink

'Getting as close as I can to the brink without actually falling over.' That's how one caver I know describes his approach to danger in caving. Just how important the element of danger is in the appeal of the sport has been covered earlier, and it is certainly true that a trip completely free from hazard would be a dull affair indeed. As one's experience underground develops, so you can tackle more demanding caves and parts of caves but with the same degree of safety as before.

The grim consequences of stepping over the brink were described in the previous chapter, cases where the person or people involved were too ambitious, or where objective danger finally caught up with them. But in caving, as in other walks of life, there are those who have stepped over that brink, yet backpedalled fast enough to come out of the situation alive. Sometimes this has been a fluke (or a stroke of good 'luck' as some would say), and at other times an example of endurance and expertise.

Howard Butler lived, thanks to the non-stop efforts of rescue teams when he suffered bad injuries down Llethrid Swallet, South Wales. Although longer than the average, this cave rescue was otherwise typical of many – with the exception that as well as fitting the victim to the cave, the cave was – to some extent – modified to fit the victim.

Butler had fallen at a point about 600 feet inside the cave, and it was obviously impossible to get a laden stretcher through certain parts of the passage. Equally, there was no hope of manoeuvring him out other than on a stretcher. As 20-year-old Butler sweated it out in the cave, tended by relays of doctors, two courses of action were started upon. Plans were made to start drilling into the cave from the hillside above, and explosives experts began to 'straighten out' the wilder convolutions of the entrance passage with small charges of gelignite. Then came the critical point: was it worth keeping Butler another day or more in the cave until the drill broke through, or should he be subjected to the rough passage out? Rescue leaders were satisfied that blasting had cleared away all really critical obstructions, and the carry-out was started. The gamble came off, and Howard Butler was dragged out of Llethrid in $2\frac{1}{2}$ hours.

Howard Butler was just one of many people who, over the years, have been pulled back from the brink of disaster by volunteer cave rescue teams. It is true that some of them have only survived the ordeal because of their exceptional strength, endurance and willingness to help their rescuers; but in these cases, it is the rescuers who save the day. On other, rarer, occasions, when no rescuers have been at hand to help, sheer determination has pulled cavers through.

Such a situation arose for two British cavers, Bill Little and Lewis Railton, in 1951. They entered the cave known as Ogof Ffynnon Ddu (nowadays the longest in the country) on a routine surveying trip of a high part of the dry Rawl Series, on the morning of Saturday 25 August. They carried out their survey, with a break for a substantial lunch (substantial in caving terms, for it included soup as well as bread, cheese, and hot coffee!), then started the return journey. This was halted when, from the top of a ten-foot climb down into the main passage, they saw a powerful torrent where there should have been a small stream.

Escape was obviously out of the question, so Little and Railton made their way to the dry Sand Passage. There they were safe from the floods, but were faced with the problem of a fairly strong draught. Whilst the draught which comes whistling under the door at home on a winter's evening is no more than a persistent nuisance, the two cavers were highly experienced and knew that if their stay was to be a long one – as seemed likely – this particular draught could be a killer, increasing the chance of exposure considerably.

The South Wales Caving Club, based virtually on the doorstep of OFD, have long followed a policy of leaving emergency rations in various parts of the cave for just such an eventuality. The two men had access to one of these depots, and, besides iron rations, found a small pick. Using this they built a small stone wall across the alcove in which they sheltered; a simple enough task, but one which fended off the worse of the draught. Their experience had also taught them something else: the danger in scooping out hollows in the sand to keep warm. Whilst such hollows do keep a stranded caver out of any air currents, they serve as traps for accumulation of mind-dulling and lethal carbon dioxide. As it was, they still noticed a tendency for this gas to build up, making them lethargic and giving them both headaches.

On Sunday morning they opened one of the emergency rations, melting down chocolate to make drinks of cocoa. They took careful stock of their supplies, and estimated that they would last them until Wednesday night. The question which nagged at them was how long would the flood last? An inspection of the water in the main passage that afternoon revealed that it was even higher than on the previous day, and they knew then that this was no mere temporary flood.

They made a number of trips to the main passage, both to leave updated notes on their progress for any rescuers who might spot their rucksack which they left in plain view, and to examine the water level. They found that they had to tear themselves away from these water-gazing sessions, and both admitted afterwards to a strong temptation to leap into the torrent and be carried through to the entrance – nothing short of suicide, they knew in cold logic.

Bill Little, who admits to having poor circulation in his limbs, calmly estimated that he would survive for from 12 to 18 hours after the fuel and food ran out. He considered making a will to tidy things up, but thought that it would not be the best thing to do in the interests of morale – particularly if he had to ask his companion to do it!

Then, on Monday afternoon, came the first sign that the flood level was dropping. Impatiently, they settled down in their alcove yet again – to risk an exit then would simply wipe out their hard-fought-for chance of survival. By 6 pm both were feeling ill with stomach aches and headaches, but these were forgotten in a trice when, an hour and a half later, they heard the approach of the first rescuers. Not only were Little and Railton able to walk out of the cave unaided, they actually lent a hand to one or two rescuers at the tricky submerged pit crossings in the main streamway. Their coldness, hunger, aches and pains became pale memories as rescuer after rescuer greeted them, and they tried in some measure to express their gratitude at the unstinting labour which had gone into diverting the water on the surface.

Exactly one year after Little and Railton had sat out their floods in OFD, the stage was set for an even longer drama – this time in the massive Swiss system of Hölloch (Hell Hole), now the second largest known cave in the world.

With three companions, Dr Alfred Bögli set off on an anticipated 24-hour surveying trip $2\frac{1}{2}$ miles inside the enormous network of passages. The weather forecast had not been particularly bad, but did foretell a deterioration in the near future. Dr Bögli bore this in mind of course, and was prepared to cut short the trip if things got too wet. What he didn't know was that during that night, while they surveyed leg after leg of passage, nearly two inches of rain fell on the surface above them, and quickly saturated the fissured limestone.

The cavers surveyed to the very end of the passage they were working in, then, at three in the morning, started on the long tiring journey out. They were soon shattered out of their drowsiness, though, for at the *Todesschlund,* where only a few drips normally break the silence, water gushed from the roof. On, ever faster, through the *Hexenkessel* and the *Regenhalle* – both resounding with clamour of the water pouring in from above. The four men were still above the flood level, and the next part of their journey would take them below it. A few silent passages gave them fresh hope . . . if only the *Wasserdom* was not flooded, they might make it yet. They reached the *Wasserdom,* and found the waterfall pounding into a lake which certainly hadn't been there on the way in. They carried on, pushing as fast as they could, pushing to the very limit, for they knew they themselves were getting very close to the brink.

Gasping for breath, occasionally stumbling, never halting . . . down the *Domgang.* Back in the *Wasserdom,* the new lake had filled the chamber, and the floodwaters were already spilling over into the passage they had taken, a swirling, foaming infant of a stream, growing rapidly to a giant testing its strength against the very walls of the cave.

Dr Bögli recognized a small stream, still at its normal size: a pinprick of hope in a more and more desperate situation. That pinprick was extinguished with the hammerblow discovery of conditions in the *Aquarium,* a foam-capped maelstrom, growing by the minute.

Escape now patently impossible, Dr Bögli did a rapid about-turn and they fled back through the flooding passages they had only just come down. The four men escaped from the trap they had found themselves in, and reached the dry chamber *Riesensaal,* exhausted but alive.

Whilst rescue teams waited outside, occasionally plumbing the water depth but finding it unvarying, Dr Bögli prepared himself and his team of younger companions for their enforced stay. Food was not too great a problem, and enough was scrounged from various depots in the cave to provide two meals per day at a bare 300 calories each for up to three weeks. Two of the cavers made a 13-hour trip to another part of the non-flooded cave to salvage a supply of carbide from a camp there.

The party, then, had the essential supplies to keep it going – at least, for three weeks. What worried Dr Bögli more was the mental state of his three companions. The doctor himself is one of Europe's most experienced cavers, and knew how to 'tune' his mind into an accord with the cave, regarding this experience as a longer than normal caving trip rather than an entombment in the bowels of the earth. Now all he had to do was convince his companions – day after day after day.

Dr Bögli adopted two lines of approach. First, the four took the opportunity of studying the behaviour of the cave's water under flood conditions, and of thoroughly exploring their immediate surroundings (without too great a drain on their lighting or energy resources, of course). Secondly, he insisted that all four resolved to keep their tempers with each other – there was enough stress in the situation without any extra being brought in.

The rules of the waiting game were observed, for each knew that his survival depended on it. It was a cold game, too, with an air temperature of about 43°F. The fourth day passed, each hour dragged out intolerably. And the fifth day. And the sixth. And the seventh.

The conversation, when it came, was almost entirely about food. Each caver in turn would describe the meal of his dreams, and Dr Bögli would smile as the others spoke of sugar and cream – he admitted later that he had to steel his nerves, for sugar and cream seemed to become the centre of the others' universe! All the time, Dr Bögli kept alert for one thing above all others, that return of air movement in their part of the cave. For when he felt that draught on his cheek, he would know that the sumps had opened again.

He waited for that draught through the eighth day, and the ninth. It came, just a whisper of air on his cheek, on the tenth day. Their wait was over. With infinite caution, they began their journey to the surface, wary of the still high water level and their own depleted energies.

The sight of a cave entrance after a particularly long and gruelling trip is a glorious vision in normal caving. What those four cavers must have felt when they saw the entrance of Hölloch, ten days after they had last passed it, cannot readily be appreciated. Nor, I suppose, can their feelings immediately afterwards when they found the entrance gate locked, and the rescuers temporarily absent from the scene! Job himself would have exhausted all patience by this time, and who can blame the four for furiously attacking that impertinent gate until, at last, they stood outside, drinking in the sweet night air.

The survival of Dr Bögli and company was never dependent upon chance. They were patient, sensible and careful. The escape of Marcel Monnayeur and Roger Jaquin was due to pure chance, and it was carelessness which put them in jeopardy in the first place.

They decided to explore the underground river of La Verna, in France, with an inflatable canoe and two acetylene lamps. Their journey into the cave started well, but they became tired of having to cope with the two hand-held lamps when they had to manhandle the canoe from one stretch of water to the next. To make things easier for themselves, they left one of the lamps behind, to be collected on the way out.

In two hours they had reached the sump which marked the end of the cave, just over half a mile from the entrance. Jubilant at the success of the trip, they started back. Almost immediately came the horrifying sound of rock tearing the canoe's fabric, and air escaping. Monnayeur and Jaquin found themselves swimming for their lives, and without any light, for their one lamp had sunk during the accident.

Jaquin tried repeatedly to recover the lamp, but his dives were useless. Frantically, Monnayeur scrabbled at tiny holds on the wall in an effort to keep their only matches out of the water. He, too, failed. The two Frenchmen fought blind panic as they struggled to keep afloat. Then Monnayeur found a crack in the rock and, wedged in this, supported his partner who bound string round the rent in the canoe.

In the blackness of the cave – the blackest black in or out of anyone's imagination – they swam for hour upon hour, kept afloat only by the remains of the canoe. They could not understand why they found no way out of the freezing water. Only later did they discover that, like the three students in Frauenmauerhöhle, they had been following the wall of a large pillar in the centre of the final chamber.

Then came the one lucky chance as their fingers, numb from immersion, encountered a ledge just clear of the water. With a final burst of energy, they hoisted themselves on to their 'desert island' – all three feet by four inches of it.

They perched precariously on that sliver of rock – no larger than a very modest mantelshelf – for 57 hours.

Those 57 hours were filled with hallucinations, as splashing water made them hear approaching rescue teams, and flashes of light danced in front of them. It seems doubtful that, if one had slipped back into the water, he could have got back again; no two men can ever have fought sleep so relentlessly. Then, 62 hours after they had entered the cave they saw the lights and heard the shouts which were not illusions. Their prayers that the innkeeper in La Verna would miss them had been answered.

Dr André Mairey, the French caver who had to face the first descent of Pierre St Martin after Loubens' fatal fall from the winch cable (see page 168), proved his remarkable courage and coolness on yet another occasion.

It was mid-afternoon on 11 November 1950. Although it had been raining for some days, the river flowing into the French cave Trou de la Creuze was at its normal level. Dr Mairey decided to proceed with his plans for studying the insect life of the cave, and set off with seven companions. As they made their way through the low passages, one of the youngest, Fréminet, found the water too cold and set off out. The other seven continued for about half a mile, to a point where the narrowness of the passage prohibited further progress, then they too started the return journey. It was then they first noticed the rising water level, slowly at first, then increasing by the minute.

Naturally, none wasted any time, and in fact Dr Mairey – who had been bringing up the rear on the way out, anyway – found himself left behind. He caught up with one man, Claude Vien, at a point where the water had almost reached the roof. Dr Mairey had no time to warn Vien, who immediately dived into the swirling floodwaters. Mairey could only hope that the others' impetuosity had carried them through to safety, and set about resolving his own highly dangerous predicament. At first he tried to retreat a short distance to where he had noticed an enlargement of the passage, but the current was too strong. He had no choice but to stay where he was.

The passage he was trapped in was barely three feet wide, and discouragingly low. But, without the luxury of choice, he chimneyed up between the walls until the roof was pressing against the back of his head. He had done all he could. The rest was up to the water . . .

Fréminet, who had left the team early in the trip, surfaced without any difficulty. Later that day, when he saw the cave river rising to a raging flood, he called for help – and help came pouring in. Fire brigades, local cavers, police, villagers, and aircraft placed at the disposal of Paris cavers by the Air Minister himself. But there was nothing they could do except, like Dr Mairey, wait.

At 2 a.m. on 12th November, Antonio Salvador and Jacques Durupt – two of the six who had dived for the surface – quit the cave. There were no cheers of jubilation, though, for both were drowned, disgorged by the swollen river.

Within the cave. Dr Mairey had been pressed against the roof, muscles straining to maintain his hold, for eight hours. The water was level with his shoulders . . .

In the first dull grey light of the morning, those waiting on the surface spotted a couple of helmets and a scarf swept out by the river. They hoped against hope that they had been deliberately thrown into the flood by the trapped men – signs that they were still alive. Whilst pneumatic drills and explosives were used to enlarge the restricted cave entrance in an effort to allow the river a more rapid exit, the first team of rescuers struggled against the current for the first sixty feet. There, in a narrow section, they found the body of Raoul Simonin.

The rescuers fought on for hour upon hour, bringing out Maurice Roth, then Claude Vien, then Michel Mozer. All were dead. The Trou de la Creuze had bitten back with a vengeance; six men were dead, battered horribly by the submerged rocks, and there seemed no chance that the cave would spare the one man yet to be found, Dr Mairey.

But that one man *was* still alive, refusing to slacken his hold against the rushing, life-sapping torrent, his face only inches above the water at its highest point. How he managed to do this for those first eight hours is a miracle. That he managed to do it for fifteen hours is incredible. When the full length of time is known, twenty-seven hours, we have gone beyond the bounds of normal human endurance. Only Dr Mairey himself can ever know and appreciate fully what those twenty-seven hours of hell meant in human terms. As a caver and a writer, I should have given anything to have witnessed that meeting between the doctor and his first rescuer . . .

What happened in the Trou de la Creuze? Why had the cave not flooded earlier in view of the many days of rain? Subterranean hydrology can be a mystery at the best of times; in this case, it is generally supposed that somewhere in the heart of the cave a natural barrier, a dam, burst under the strain of the accumulated water.

Luck played no part in Dr Mairey's survival, but it must have been working overtime for the astonishing escape of John Stevens – the most remarkable caving episode in recent years.

Sixteen-year-old Junior Private Stevens led a squad of other young soldiers into Derbyshire's well-known Carlswark Cavern – a system graded only Difficult – in July 1965. He came across a sump, and, being unsure of the way yet a 'strong swimmer', dived happily into its depths. While unrestricted sumps of five, ten or even twenty feet are being commonly free-dived today, Stevens was taking on rather more than that. A sump of 200 feet, in fact.

The Derbyshire rescue teams got on with their job of pumping out the sump – possible in this particular case, although not in many others – but without much enthusiasm for their task of recovering the body. The sump was broken, after some twelve hours, and the first team waded cautiously in.

They found John Stevens, shivering but very much alive, in an air pocket – 100 feet inside the sump passage!

Stevens, who told the reporters waiting outside the cave that he had 'never really been scared', said that he had fallen asleep several times during his imprisonment. He explained that he had been wearing only a pullover, trousers, beret and boots.

That Stevens had remarkable free-diving ability is beyond question, but, without appearing ungracious, was he really the 'hero' of the rescue as the newspapers maintained in their banner headlines? To many cavers, heroism did not come into the affair, but rather an especially generous serving of pure unadulterated good luck.

Chapter 10

Experiments in Time

Companionship born of many long hours of shared hardship and hazards is often one of the reasons cited by cavers for the pursuit of their sport. While those friendships in ordinary life may well be strong and long-lasting, those forged between cavers who can count on their companions to get them out of any trouble – and who can in turn be counted upon to give their own help – are of a particularly strong nature. Besides the additional security provided by two or three companions underground, there is the opportunity to share the awe and the wonder, to be able to point out to someone the tiny gem of a formation hitherto undiscovered in some overlooked pocket or new passage. The essence of being one of a team is that pleasures are doubled and worries halved.

How, then, can one possibly explain the motives of that uncommon breed of man – or woman – who is not only prepared but apparently quite happy to spend days, weeks, or even months in complete solitude deep in some cave?

Though the number of people who have made long solo underground stays can hardly be called a spate, such events do catch the attention of the world's Press – and, consequently, the public. The story has not only the automatic spice added by any headline with 'cave' in it, but the additional one of a man facing the dark, that age-old horror of all races, all by himself. To the man on the surface it is always the darkness of caves which, in his mind, constitutes the greatest adversary of all; not the cold or the damp or the flood risk or the danger of a collapse, just that perpetual night.

Without in any way belittling the efforts of those people who have chosen to face long periods of solitude underground, it is important to give the general opinion of them held by the majority of cavers. It might be thought that their exploits would have elevated them to the status of supermen in their sport, and that their enterprises would rank as the greatest in subterranean endeavour. Far from it. The truth is that many cavers, whilst acknowledging the undoubted staying power and patience of these men, have strong doubts as to whether they contribute all that much to the sport or its science. If solitude and darkness are the prime requisites of such enterprises, then why not conduct them in some old concrete bomb shelter or bunker? Given thick enough walls and doors the isolation would be to the same degree, and the problems of supply and communication would be infinitely less.

Possibly the main reason for the choice of caves in solitude experiments is that the participant – or victim – knows full well that failure to stay the required length of time is far less likely in a cave. Given only a couple of iron doors between him and the sunlight, the experimenter would have to fight the temptation to simply open them to get his freedom, particularly in his moments of low ebb. Inside a cave, the deeper or farther in the better, he knows that 'escape' is more difficult – particularly if the surface support party has removed the ladders from shafts between him and the surface! Another important reason for choosing a cave in preference to a bunker is that while squatting in a bunker would, understandably, be classified by the Press in the same category as sitting on top of a pole, the same length of time spent squatting in a cave would be an 'adventure', and far more respectable. Absent from his normal occupation for weeks or months at a time, he depends on financial support from the media for his story. In effect, he has no choice: he must go for the added glamour of caving.

The story of grin-and-bear-it underground sojourns started in 1953, when wiry Geoffrey Workman clambered down to a quiet chamber in Gaping Ghyll, the Yorkshire pothole, and there set up camp. Geoff, who admits to his inspiration being fired by his reading of Verne's *A Journey to the Centre of the Earth* as a schoolboy, stayed underground for two weeks – modest by today's standards, and his own later records, but hailed then as something not far short of a miracle. And even that stay was without perfect solitude, for a small party visited him at the end of the first week, bought a spare bulb from him, then wandered off to see the rest of the cave. With no special difficulty – except for a still unsolved deep rumbling which brought him rushing from the tent – Workman completed his stay and was taken immediately to a television studio where he completely baffled the panel of the then popular television game, 'What's My Line?'. And hardly surprisingly.

One of the best documented attempts took place in a pothole of the French Alps, the ice-lined Scarasson Cave. The French caver Michel Siffre, then only 23, was intrigued by the glacier which had been found in the system at a depth of 375 feet, and believed that only a prolonged stay would enable him to examine it thoroughly. He was also intrigued with the prospect of increasing man's knowledge of human existence in an adverse subterranean environment. Besides the usual high humidity, Scarasson also offered the additional discomfort of a temperature constantly at, or even below, freezing point.

Besides straightforward scientific curiosity, Siffre – like Workman later on – was itching to know just how well and how completely a fully 'civilized' human being could take to living in the caves again. He commented cynically that at the present rate of progress we may well all end up there again. On the purely scientific side, Siffre's experiment was watched by experts with considerable interest: how would his body react to the absence of a daily cycle established by the rising and setting of the sun? When he realized that only by having no reference to time 'outside' would he be able to properly study his body's time-cycle, Siffre made the rather brave decision to go down without a watch: brave, because we are all slaves of the clock, despite our protestations at being tied to a timetable, and feel a peculiar insecurity when subjected to true timelessness.

Siffre's first problem was to find backing for his experiment, and he found that his age and relative lack of reputation were very much against him (he was an undergraduate at the Sorbonne, and his increasing absences were indulgently overlooked as those of an *enfant terrible*). Finally, he met the climbing hero of Annapurna, Maurice Herzog, a high commissioner for the French youth and sports programme, and convinced him. With Herzog's patronage, Siffre was able to start the long process of gathering food, clothing and equipment for his long cold stay. At that time there was no institute in Paris devoted to speleology, so Siffre promptly founded one; letterheading of the 'Institut Français de Spéléologie' brought an agreement from the Centre for the Study and

Research of Aeronautic Medicine to put Siffre through a series of physical tests before and after the experiment, and from the Cité Universitaire Clinic to analyse samples of blood, urine and so on. With the agreement of the Compagnie Républicaine de Sécurité to provide men from their mountain section to give assistance throughout, and constantly man the surface telephone, and a promise of helicopter transport for the heavy supplies up the mountain by the Protection Civile in the Alpes-Maritimes, Siffre was ready. A final blessing from Dr Alain Bombard – whose remarkable experiment in 1952 showed how men can survive at sea in a dinghy – made his day.

Helped by other cavers, Siffre set up his camp on the glacier, with the 8-foot wide and 13-foot long tent as a centrepiece. A final round of farewells, and the helpers made the ascent up the shaft, the noise of their departure getting fainter and fainter. The Frenchman was alone, the frail telephone wire his only link with the surface, and he felt overwhelming and suffocating fear . . .

The surface support party had been given explicit written instructions that no one was to enter the cave for at least one month, under any circumstances. Just how far they would have stuck to this request if things had gone very wrong is difficult to say. As it was, Michel Siffre had to brace himself for at least a month-long solitude, in the paralysingly cold atmosphere of the cave, and in the event stretched his self-imposed purgatory to 63 days.

He attempted to keep a full diary, but found that his memory failed him constantly. Often he was unable to remember what he had been doing only minutes before. He adopted the routine of writing notes down after anything of significance had happened; his diary also differed in that dates and days had no real meaning, so he used headings of such-and-such an 'Awakening'.

In his diary, Siffre is quite open about the gradual mental and physical deterioration which set in, and admits to moments of excruciating fear when the chamber in which he lived resounded to the crashing of rock and ice falls. Whilst an inactive or dead cave changes very little over tens of thousands of years, rock coated with ice is subjected to a constant gnawing – something Siffre grew only too aware of.

One of his worst problems was that of damp, and the floor of his tent was usually swimming with a thin layer of water. This led to a demoralizing and constant battle to keep his clothes and foot-wear dry, and the only time his existence was anything like com-fortable was when he took to his sleeping bag. Fortunately, he had chosen to use a folding camp bed rather than an inflatable mattress; the latter would have resulted in a wet sleeping bag, and without that ultimate retreat it is doubtful that Siffre, despite all his deter-mination, could have endured the whole stay.

'Desert Island Discs', the popular radio programme, asks which eight gramophone records one would choose to take on an island exile, and it would seem reasonable to assume that under such cir-cumstances music would indeed be a balm to the soul. Siffre, a lover of fine music, found after a while that music meant nothing to him; even his favourite records became meaningless cacophonous jumbles.

The chief requirement of the surface support party, at the other end of the telephone wire which snaked up from Siffre's canvas home, was that they should never give the slightest indication to him of the real time. This was a difficult, and sometimes exasperat-ing imposition, for if one of the experimenter's calls disturbed them in the middle of the night, voices had to sound bright and cheerful despite their having been dragged from slumber. On one particularly taxing day, Siffre woke them up at 5.50 a.m. to tell them he was starting his breakfast, after which they went back to bed. At 7 a.m., just over an hour later, he woke them again to inform them that his 'day' was over and he was going to bed. At 9 p.m. the telephone jangled again: Siffre was up and wishing them a good morning – when they had just retired. The hectic interlude for the ever-patient surface team was capped when Siffre put through his next breakfast call . . . at 3 a.m.!

Away from the perpetual cycle of night and day as governed by the sun, Siffre sank into his own time zone. His estimations of his working or waking hours were utterly wrong: a 'day' which he estimated at only seven hours actually occupied him for nearly fifteen hours. Despite the often erratic lengths of sleeping and waking periods, Siffre's experiment showed that in fact he con-formed very nearly to a 24-hour day; he should have trusted in his combined sleeping and active periods to add up to 24 hours.

Only too aware of his vulnerability in the event of an accident, Siffre made cautious excursions into other parts of the cave, exploring the limited range of the system, and taking ice and geological specimens. The whole affair nearly ended in disaster on only the fourth day as he clambered about the ice slope trying to position himself for some photography. Siffre lost his balance and started sliding down the ice to the edge of a small shaft – he was saved by his free hand grabbing a metal peg just a few feet from the lip. After that he resolved never to venture on the ice again without a pair of crampons (grids of steel spikes) on his boots.

Siffre's diary reveals his growing obsession with the minor details of food and comfort; survival became the key thought in his mind. After two or three daily meals early on in the experiment, he soon became satisfied with just one main meal, remarkable considering the true length of his working period. Attempts at dishwashing after each meal soon gave way to an unabashed table routine of eating straight from the pans.

As the experiment progressed, Siffre began to suffer from visual hallucinations as well as a worsening strabismus, or cock-eyed squint. His colour perception was altered, too, and tests revealed that he saw green as blue – a condition which persisted on his return to the surface. Attacks of dizziness became fairly frequent on his forays outside the tent for provisions, and over the weeks a fear of a damaging fall kept him confined for most of the time to his tent. Despite half-hearted attempts at clearing up his 'home', the pile of rubbish outside the door often reached thigh level. Fortunately the freezing temperature slowed down the process of decay in the discarded food.

Siffre's diary, published in his book *Beyond Time* (Chatto & Windus, 1965), gives some fascinating insights into his state of mind, as the following extracts indicate.

35th Awakening (9th August): '. . . I am becoming absent-minded. This morning I went out to answer a call of nature and upon reaching the foot of the 130-foot pothole, I couldn't for the life of me remember why I had gone out! '

41st Awakening: 'After the stupendous cave-in of August fifth, I wrote a letter . . . which was my last will and testament. I am becoming more and more nervous, but I'd rather die than call for help . . . The disorder in the tent is indescribable . . . I have suffered more here than I have admitted in this diary, for the cold is terrible, particularly when combined with the humidity. . . . '

9. Double lifelining solves the problem of how a caver, having lifelined his
companions down a pitch, can himself descend on the security of a rope;
the rope runs through a pulley left at the top. Here the technique is being
used in New Goyden Pot, Yorkshire

10. A blanket of foam caps the sump in Derbyshire's P8 cave; from here on the divers take over

Camp One of a British expedition 1,200 feet down Italy's Antro della Corchia. Tents are used to give psychological rather than physical comfort in underground camps like these

44th Awakening (14th August): 'I lay for some time in my sleeping bag, alone in the vast silence of the subterranean night, listening to some Beethoven sonatas. The effect was fantastic. Impossible to count short lapses of time by the records; the beginning and the end of a record blend and become integrated in the flood of time ... Anyway, what does it matter? Time no longer has any meaning for me. I am detached from it, I live outside time ... At the beginning, I really suffered, but now things are much better. I can even say, life is great!'

50th Awakening: 'Singing is a great comfort. You hear your voice as if it were another self; you have the impression of a human presence. But imagine the weird scene: a man alone in the depths of the earth singing aloud in the darkness and cold, a captive within walls of rock ...'

51st Awakening (17th August): '... And I am too young to end up a shapeless corpse smashed between two boulders in the cold and darkness, and the silence! This total silence is horrifying; there is only the drip-drip of water falling. Sometimes I *listen* to the frightful silence, and my mind reels. Sometimes I could weep. Have I wept? I don't think so; yet during the enormous cave-ins tears start to my eyes. I now understand why in their myths people have always situated hell underground ...'

Michel Siffre's own private hell came to an end on 14th September 1962. His own time graph told him that it was only 20th August; in his experiment beyond time, he had 'lost' 25 days.

Siffre's return to the daylight was a severe trial for him. He was hauled bodily up the entrance shafts, often in tears, the ascent punctuated by periods of complete blacking out. The Press and other media greeted like a conquering hero this new-style time traveller.

Although Siffre claimed that he was in no way interested in the fact that he had established some kind of world record, a world record it was – and as such, stood waiting to be broken. The breaking came rather sooner than anyone might have expected when Bill Penman, an Australian caver, emerged after 64 days underground; he was forced to surface due to general deterioration and failing eyesight.

Back in Britain, Geoff Workman was watching these developments with interest, and a growing conviction that two months was not the limit for such experiments, as was being maintained in many quarters. He thought that careful planning and special techniques were the answer and decided to put them to the test.

Without the vast choice of caves that France offers, Workman had to find one which had a locked door or was permanently guarded. This led him to search for a suitable non-tourist extension to a tourist cave, and he found just the place he needed in Stump Cross Caverns near Pateley Bridge, Yorkshire. On the understanding that he made telephone contact with the cave owners once a day – and mindful of the added attraction of an in-residence caveman – the management agreed. Working to a budget of nearly £1,000, and with half a ton of equipment to see him through, Workman settled down to his long wait securely sealed from the prying public by two locked doors in the approach passages.

Apart from the usual pre- and post-experiment tests, Workman collected urine samples each week, poking them through a small hole cut in one of the doors for swift analysis. Other regular measurements were made of pulse, temperature, general fitness, muscular strength, intelligence, colour perception, eyesight, hearing, sleep variations and physiological variations.

Workman sited his camp at the top of a sandbank in a dry gallery, a position which kept him clear of the dangers of flooding and CO_2 accumulation. There were significant differences between his programme and that of Siffre: the cave temperature was much more bearable, there was not the problem of equipment being constantly soaked, and Workman did take a watch with him. But despite his early attempts to work to a fairly normal schedule of 24 hours, going to bed at 10.30 p.m., he found himself taking longer to fall asleep each night and was waking up later each morning. Eventually he gave up working to the clock, with its rather artificial pace making in the absence of daylight, and slept and got up when he felt like it. His sleep rhythm was being forced on to him by some internal working of the body, over which he had no conscious control. During the period of twelve weeks in which he slept according to need, Workman's 'day' averaged $24\frac{3}{4}$ hours.

Workman cracked the two-month endurance barrier without difficulty, emerging after 105 days underground, and obviously suffered far less from the enterprise than either Siffre or Penman.

Just as some scientists had once predicted that man could never withstand the rigours of space flight, and have been proved very largely wrong, so our solo cave squatters have disproved the doubts expressed about the human endurance barrier in underground solitude. Since the first attempts by Workman and Siffre, the records have been broken repeatedly, with the participants in various degrees of health and sanity at the end of each effort.

For a fee of £5 a day from the owners of a Cheddar cave, David Lafferty completed a stay of 127 days. Filling in his time with eating, reading and sleeping, and a weekly game of darts, Lafferty lost almost a month by his reckoning. He emerged to fulfil his desire for a pint of beer, and baffle the scientists with his alarmingly good health.

Now that the 'endurance' barrier has been proved to be something of a myth, Siffre and his counterparts are having to find ever newer approaches to their exploits. In the Frenchman's longest stay at the time of writing, 205 days in a Texas cave with only six mice for company, his object was to discover whether he could slip into a 48-hour sleep/activity cycle; this he did briefly on just two occasions.

Most of the men (and the very few women) who have made their reputation through these chilly epics are indeed cavers to start off with. But it is debatable whether their efforts have very much to do with either the sport or the science of caving. Their choice of caves for incarceration would, it seems, be more a matter of convenience than anything else.

As nothing is new under the sun so, it would seem, is the case under the ground. Tucked away in *Celebrated American Caverns*, a classic of 1882, is an often overlooked reference to some strange stone cottages in Mammoth Cave, Kentucky. Built in 1843, these were used to house 'fifteen consumptive patients, who here took up their abode, induced to do so by the uniformity of the temperature, and the highly oxygenated air of the cave, which has the purity without the rarity of the air at high altitudes.' In one of these dwellings, a desperately sick man by the name of Mitchell sat it out for no less than five months before dying. Not a bad effort for a man far removed from the health and vigour of our Workmans and Siffres!

Chapter 11

The Golden Dream

Every sport has its golden dream. Many are realized, such as the
four-minute mile, the ascent of Everest, and the passage round the
world by a solo yachtsman. It is an interesting phenomenon that
there is never a shortage of new dreams either: with Everest's
summit reached, mountaineers are now making attempt after
attempt on a direct ascent of the South-West face of that mountain, and solo passages round the world must now be against the
prevailing winds, or via the North-West Passage, or with one leg
tied behind one's back.

American cavers have had their golden dream, too. As the adjacent Kentucky systems of Mammoth Cave and Flint Ridge Cave,
like Topsy, just grew and grew, they began to entertain the fanci-
ful idea that one day, just one day, the two might be connected.

Mammoth Cave was discovered at the very beginning of the last
century, the exact date being in dispute. The cave became widely
known for the vast quantities of saltpetre it contained, and during
the war of 1812 tons of this valuable chemical were mined there for
the manufacturing of gunpowder. Naturally, Mammoth was soon
opened for tourists, but the owners of the section reached by what
is now called the Historic Entrance found themselves with a fight
on their hands.

George Morrison was a particularly determined man – and a
shrewd one. He reckoned, as others have done, that Mammoth
must surely extend beyond the boundaries of the Cave Management, and located some of these extensions by exploratory drilling.
He hired a gang of labourers, and within two months had opened
up another entrance to Mammoth, now known as the Cox entrance
for the land there was Cox property. However, Morrison was
rather taken aback when the owners of the Cox property prevented
him from exploiting the new entrance, but he did take the chance

of spending a fortnight surveying as much of Mammoth as he could.

Morrison spent the next five years buying up or leasing free land round the Mammoth Cave Management property, then, through his Mammoth Cave Development Company, opened up his section of the system. Why bother going through the old smoke-covered section of Mammoth, proclaimed his publicity handouts and posters, when you can see this wonderful new section of the same famous cave? To really rub it in, Morrison's guides would politely invite tourists who had already entered via the old entrance to exit, quite free of charge, through *their* cave. The subterranean feud ended in a court battle, which Morrison won when it was decided that the name Mammoth Cave was fairly applicable to all of the system, and not just that part reached by the old entrance. The whole area around Mammoth began to sprout show caves over the years, and there was a long period of bitter rivalry in which competing businesses employed 'cappers' to reroute tourists headed for Mammoth to some other cave. The cappers would deck themselves out in official-looking uniforms, and flag down tourists' cars. There was an unwritten law against out-and-out lying, but just about everything else was accepted and the stories contrived by the cappers were great works of persuasion of which Dale Carnegie himself would have approved. Early in the day, tourists would be told of the exhausting nature of the day-long trip in Mammoth; in the afternoon, late-comers were told that really they had missed the best trip of the day, which was the all-day one! The slightest hint of rain brought out bloodcurdling tales of Mammoth floods which omitted to include the date of many years before.

Bitterness rose to a climax one morning when the owners of Crystal Cave found that the embalmed body of Floyd Collins had been stolen from the coffin within the cave. The corpse was a chief tourist 'attraction' for Crystal Cave, and an immediate hunt was started with the aid of bloodhounds. Collins's battered body was found at the bottom of a cliff. Finally, against this troubled background, the number of tourists visiting Mammoth Cave, and those of the surrounding area, became so great that Mammoth was declared the country's 26th National Park.

Much of the exploration of Mammoth Cave has been done by a

succession of guides, from coloured Stephen Bishop in the middle of the last century to Carl Hanson, and his son Pete, just before the last war. It was Carl Hanson who first glimpsed the glorious gypsum formations in New Discovery, and whose son's initials were to prove so important to an exploratory party in 1972.

Situated about a mile and a half north-east of Mammoth Cave Ridge, on the same large plateau bordered to the west by Green River, lies a massive sandstone-capped limestone block – Flint Ridge. For years, the name has caused cavers' muddy ears to prick up at its mention, for in these few square miles have been concentrated tens of thousands of man-hours of highly intensive cave exploration and surveying.

The fascinating story of the discovery of the huge Flint Ridge cave as it is known today is a complex one, for the Flint Ridge system has been pieced together slowly, an amalgamation of a number of already known caves, like the stringing of a necklace one pearl at a time.

Salts Cave was known to the earliest settlers, and to the Indians, many generations before them, who mined gypsum there. Unknown Cave came next, but it is doubtful that early cavers explored much beyond the shafts and collapse areas terminating the entrance passages – except possibly for one or two of the bravest, for a map dated 1903 showed Unknown Cave as an entrance to Salts. Through the following decades, the map – and the reputed connection – were forgotten.

At the end of the last century, Colossal Cave's marvels were discovered for the first time – possibly by a Negro guide from Mammoth, William Garvin. Garvin found a huge corridor two miles long, its walls covered in gypsum crystals, sporting a 75-foot high stalagmite. Later, a tight connecting link was discovered with the small Bed Quilt Cave.

The year 1917 marked the discovery of Crystal Cave by that ill-fated explorer Floyd Collins, whose name became a household word a few years later when he was trapped by the leg in Sand Cave. During the years until his death, Collins explored mile after mile of Crystal Cave, but after the tragic episode in Sand Cave, exploration in Crystal was reduced to the odd visit, and there was no co-ordination of work.

In the early days of Flint Ridge exploration, talk was rife of connecting the separate caves into one big system, possibly even large enough to compare with nearby Mammoth. But this idea lost favour as the hoped-for breakthroughs did not appear, and pessimists pointed out that the caves of Flint Ridge were all situated on spurs of limestone sticking out from the main ridge, and there were deep dividing valleys between each of these – Mammoth Cave, on the other hand, was situated in one single mass of rock.

Co-ordination of effort is vital in the pushing of a really vast cave otherwise team after team might cover a section of passages and not know whether they are in virgin ground or well-known territory. The co-ordination at Flint Ridge came into being in the mid to late 'forties when Jim Dyer, the manager of commercialized Crystal Cave, started systematic exploration. Besides Crystal, the other major caves were examined, too, with the agreement of the National Park Service. As the cavers pushed farther and farther, they met a problem which had never really cropped up in American caving – the endurance barrier. However many hours they spent underground, it made no difference: the point was reached were it took half the available time simply to reach the previous limit.

The turning point was C-3 – the week-long expedition mounted by the National Speleological Society in 1954. During this week, 64 American cavers threw everything they had at Crystal Cave's inner regions; although finding no outstanding extensions, the effort did spark off a ceaseless attack on Flint Ridge. It was realized that large parties only compounded the difficulties so teams of half a dozen or less became the norm. Slowly, stockpiles of food were accumulated at the limits of exploration in Crystal, and the other caves of the Ridge came in for their share of probing, too. Just as momentum was building up, the explorers ran into the most damnable obstacle of all – Red Tape, yards upon yards of the stuff. The National Park Service imposed severe restrictions on caving within their domain, and patrols were made of the caves themselves.

Where there's a will, or better still unexplored cave . . . Explorations continued, on a reduced scale obviously, but enthusiasm was unabated. Discoveries were now kept secret when they were made under National Park territory, such as the exciting finds made in

Unknown Cave when those shafts which marked the former end were passed. As if rewarding the cavers for their perseverance in the face of officialdom, the finds were on a grand scale. One day's pushing in Unknown Cave revealed no fewer than five miles of new passage, and the forgotten ambitions of linking all the Flint Ridge caves were fired again, for passages were leading out from Unknown to Salts, Colossal and Crystal Cave itself! Quietly, work in Crystal was swung towards where Unknown Cave lay, and link after link was added to the chain: Overlook Pit; the floodable Storm Sewer; the Eyeless Fish Trail with its spirit-raising major river; Pohl Avenue, reached by a risky climb up from the Eyeless Fish Trail. Then that day in October 1955, with the climbing of another shaft, the breakthrough – the connection with Unknown Cave. Flint Ridge cave took its place at the top of the list, the longest cave in the world.

The challenge was now so great that Flint Ridge cavers took the step of launching an organization independent of the NSS, the Cave Research Foundation, with the chief objective of exploring and surveying Flint Ridge. The CRF was able to negotiate with the National Park Service the question of access, and open exploration began again.

In 1960 a three-man team led by Jack Lehrberger decided to examine a river passage in Colossal Cave. As the leader clambered up to a drier section, he had the distinct feeling of having been there before, and indeed he had, for he suddenly recognized that the passage he stood in was one leading from Indian Avenue in Salts Cave! A year later, a two-man plus one girl team moved a heavy rock obstructing progress in the Lower Crouchway area of the Crystal–Unknown complex, crawled through and found themselves with the happy job of hurrying out to announce that it was now the Crystal–Unknown–Salts–Colossal complex.

Year by year, the number of miles of cave surveyed in Flint Ridge added to the already impressive figure. By 1972, the CRF had surveyed the now integrated network of passages in Flint Ridge to an incredible 86·5 miles, and nearby Mammoth Cave stood third on the world list at 57·9 miles.

Flint Ridge was the longest – comfortably so over the Hölloch, the Swiss 72-mile-long giant – but the golden dream was there: Flint Ridge Cave and Mammoth Cave were so close. Was it an

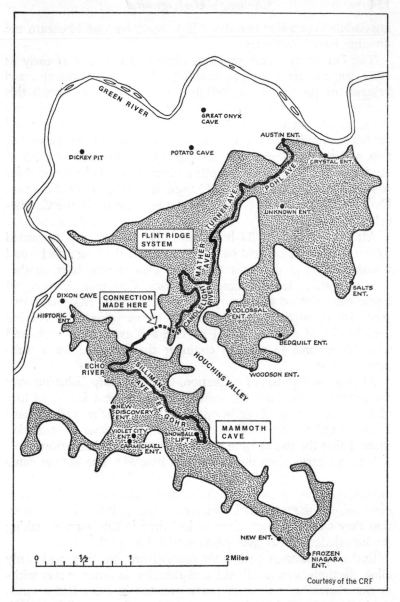

The Flint Ridge–Mammoth Cave system. The thick line shows the route of the link-up team. Shading indicates the extent of the 150 miles of passage

impossible dream that one day a link would be found beneath the dividing Houchins Valley?

The CRF cavers came precious close to an answer as early as 1964, when a survey team pushed to the limits in a mile-long crawl beyond the promising Candlelight River area. Even though this discovery pierced the rock beneath Houchins Valley to the very flanks of Mammoth Cave Ridge, further attempts were few and far between owing to the very strenuous nature of this particular trip, and the knowledge that the passage was blocked by a fall of sandstone boulders. Such tempting plums can only be ignored for so long, though, and early in 1972 the chief cartographer of the CRF, Dr John Wilcox, renewed the foundation's efforts in the Candlelight River area.

After a number of 24-hour trips, Dr Wilcox was rewarded by the discovery of a tight passage leading off the long crawl – and it was heading in the right direction. Two months later another surveying team, led this time by Patricia Crowther, 29-year-old housewife and computer programmer from Massachusetts, pushed on through Candlelight River to the new passage. Surveying leg after leg of the constricted passage was exhausting work, and when they came to a small chamber they stopped for a well-deserved rest.

After a few minutes of relaxation, their eyes idly flickering over the muddy walls, came the most electric moment in the history of Kentucky caving. Together almost, Tom Brucker and Richard Zopf noticed scratch marks on the wall, and closer examination showed that the marks were the name PETE H and an arrow.

Who on earth, they wondered was Pete H? The answer came in a heart-stopping flash – it could only be Pete Hanson, a Mammoth Cave guide of the late 'thirties, and that arrow – which would have pointed the way out for Hanson – was pointing in the direction they were heading! Hanson had died in the last war, taking the knowledge of this river passage with him to the grave.

Excitedly the team pushed on downstream, but they could only allow themselves an hour's extra exploration in order to stay within the allotted trip time. A party overdue in Flint Ridge automatically sets search and rescue procedures in operation on such a large scale that even a find of this importance could not be pushed regardless of time. The hour passed all too quickly by the curious time-warping effect a cave can have when things get interesting, and Patricia's team had to turn back, a shattering five-mile return trip in front of them. Even had time allowed, their exhaustion was so great that they would probably not have got much farther in any case.

Hardly able to contain their excitement, a group of cavers gathered on Flint Ridge on 9th September 1972. Dr Wilcox led them into the cave, via the Austin Entrance. They reached their previous point of exploration, marked by their initials on the muddy wall, and started surveying it. The temptation to leave the surveying for just this once, to press on into the unknown, must have been enormous – but the CRF has a reputation for a methodical and accurate approach to cave exploration, impossible without consistent surveying of new territory. They reached a spot where the roof dipped down almost to the surface of the water. To duck through would mean total submersion, at least of the body, and that meant a cold and shivering conclusion to the trip. There was little doubt as to what Wilcox would do. He ducked under the rocky arch.

Without warning, Wilcox found himself chest deep in a large chamber. There was a moment's unbelieving pause, then he yelled back: 'I can see a tourist trail!' The others plunged through, and their shouts of delight echoed in the chamber. Their lights revealed the glint of a steel handrail on the other side of the water – an old tourist path running through Mammoth's Cascade Hall. Flint Ridge and Mammoth Cave were connected – the golden dream was 24 carat reality.

After the jubilation, the cavers steeled themselves for the long journey back. Their final surprise was yet to come, though, for one of the team – Ranger Cleve Pinnix – had the key to one of Mammoth's entrances, and they were able to exit via the Snowball Dining Room lift, making the first through trip and still having enough time for a nocturnal champagne celebration.

For Patricia Crowther, sleep was out of the question, though, until one final loose end had been tied in. Carefully she punched the survey data into the computer terminal which resides in a corner of her home. Data assimilated, the computer plotter drew the completed map. *Then* it was time for sleep, knowing that the Flint Ridge–Mammoth Cave system was officially logged at a mind-boggling 144·4 miles of surveyed passage.

Chapter 12

Deep

Under certain conditions, skin-divers face a hazard of gas narcosis known as the rapture of the depths. Cavers have their own brand of rapture which comes with depth, but far from being a danger it acts as a constant spur, forcing cavers to push themselves to the very limits in order to gain just a few more feet of depth.

The sensation of depth in a cave is a powerful but fickle thing. Some caves, because of a gentle slope over a very long stretch of passage, reach considerable true depths (caving depth is the difference in altitude between the highest entrance and the lowest point, not the distance from that point directly up to the surface). And yet, the feeling of depth can be quite insignificant in them, for it has been gained too easily to be appreciated.

There are deeper systems in Britain than Penyghent Pot, burrowing into the flanks of Penyghent mountain above the clustered cottages of Horton-in-Ribblesdale, but there are no caves – as opposed to potholes – in which the sensation of depth is so powerful.

Wading waist-deep along the brooding river on the way to the last section, abruptly full-stopped by a deep and sudden sump, the caver . . . no; the potholer, for no other name is truly accurate in Penyghent . . . proceeds in silence, almost reverence. The black walls, the black waters, the back-of-the-mind knowledge of that endless series of pitches between himself and the sun – or will it be the stars by now? – add up to a sharp awareness of piercing the heart of the rock, of being admitted, tolerated, for a short while in a mountain's heart. And the river is the lifeblood pumping through that heart.

If one were to ask a cross-section of cavers if they would prefer to add a mile to the known length of Flint Ridge–Mammoth Cave or a mere hundred feet to the depth of the world's deepest system,

I am confident that the vast majority would confess readily to the latter preference. So, in the peculiar convolutions of the brain which reside inside the skull of homo sapiens the caver, sheer distance is not the chief factor.

Why *is* depth so precious to us? Why would the news of a new depth record bring a greater surge of excitement than news of a proportionately greater length record?

Like the question, 'Why do you cave?', it is extremely difficult to get to the kernel of the truth, but I believe it has a lot to do with the greater degree of dedication needed to gain great depths. In terms of safety – which must lie at the back of every caver's mind in considering such things – progress down a series of shafts seals the escape to safety rather more securely than long relatively level passages. Every hundred feet of level passage traversed means simply another hundred feet of stretcher carrying or dragging if things go wrong. A hundred feet of water-lashed shaft means a difficult time-consuming problem of haulage – and, as has been pointed out earlier, time is the greatest enemy of an injured caver. Tired limbs can be coaxed on the way out of a long cave; the body can be pushed and pushed to the point of collapse – but then the entrance usually comes before that point is reached. It is infinitely more difficult to find the hidden resources of energy needed to climb a ladder, or prusik up a rope, when one is at the same point of tiredness down a pothole.

To understand this is to appreciate more readily why depth has such an attraction. To push our longest caves is no easy thing, but the ultimate committal comes with the decision to reach and extend those deepest natural cavities under the earth.

A few miles outside the city of Grenoble in south-east France lies Sassenage, an old town famed for its centuries-known tourist caves, the Vats of Sassenage. It was in this lyrically named system that a handful of keen young cavers, mainly from Grenoble, learned the basics of their chosen sport, often during midnight escapades when they managed to escape the slumbering attentions of their parents.

This was just after the last war, when caving in France was finding its feet once more after long years of stagnation, and sights were aimed higher – or, more correctly, lower – than ever before.

The young Grenoble cavers had mastered their sport, and much of the Vats of Sassenage. Now they wanted to discover the source of the Germe river, which at one point burst into the Vats in a roaring flood. They turned their attentions to the vast rift-riven Sornin Plateau high above Sassenage, an expanse of limestone which resembles a sheet of white candy after the knife has criss-crossed it, dividing it into none-too-regular lumps. The only difference is that in places the point of the knife has slipped, driving deep into the candy mountain.

The French cavers started their painstaking search of the plateau in 1951, but due to filming commitments in the Vats (the result being the now classic *The Starless River*) they were unable to make a really concerted effort until 1953. Then, during a Whitsuntide meeting of the group, came the first episode in one of caving's longest and most exciting developments.

The plateau presented the cavers with holes boring into the limestone at almost every single step, in any direction. Most ended after only a short distance, but stones thrown into others clattered and rattled down for an encouraging number of seconds. A half-dozen of the most promising were earmarked for investigation by teams with ladders and ropes.

The exploratory descents were just about to begin when there came an excited cry from one of the team still wandering at will over the rock, Jo Berger. He had found a really promising hole, and efforts were immediately switched to plumbing this. A free-climb of 15 feet led to a ladder pitch of 25 feet or so. Jo Berger clambered down. No doubt he was braced for disappointment, the exploratory caver's normal return for his efforts, but this time he was turning the key to the glories of a system which now bears his name in honour: the Gouffre Berger.

And what a name! Since the mid-fifties, the (always) casual reference to one's first-hand experiences in the Berger has been the ultimate in name dropping, and guaranteed to create a momentary pause in the conversation if not exactly an awe-inspired hush – cavers are far too cynical a breed to allow any one of their tribe to balance on the glory pedestal for more than a few moments at a time.

Back to Jo Berger, delighted at his find, and announcing after his first reconnaisance that this was a 'whole little system of caves to look forward to', surely the most modest appraisal of a new find with such potential yet. The night after the discovery, the French cavers found it difficult to sleep in anticipation of what they might find next day. Breakfast was unusually early.

By early afternoon the exploratory team of Berger, Ruiz, Gontard and Jouffrey were all underground. His lamp exhausted from the previous day's excursions, Jean Cadoux waited on the surface, allowing his imagination to run with his team mates in the unexplored halls and corridors beneath the baking limestone.

Three hours later shouts from below warned him of the returning party; he lifelined up Berger, who gleefully told him of the considerable progress they had made down two major shafts and 60 feet down a third before running out of ladder. From the shaft below came a sudden horrible cry, cutting off Berger's narrative in a trice. Someone had fallen.

It was Ruiz, a Spanish student at Grenoble's Polytechnic, and a popular member of the team. Cadoux rushed down to investigate, horrified at the silence which followed the first cry. He peered over the edge of the 100-foot shaft – there, dangling not far below, swayed the ends of the ladder, frayed steel strands, with tiny gleams of bright metal where the rock had sheered almost completely through the wires. The weight of Ruiz had snapped the few remaining strands and he had fallen down much of the shaft. The lifeline by which he was secured was a thin nylon one, and the caver Bouvet had been unable to do more than check the speed of the fall even though the rope had bitten deep into the flesh of his hands and arm.

As Cadoux ran over the limestone plateau to fetch extra help, it crossed his mind over and over again: was the price of the discovery of this new pothole to be the life of his friend? Cavers took risks as a matter of course, but no cave on earth was worth that price. Happily, no such sacrifice was necessary on this occasion. Ruiz was hauled out of the cave, on the point of unconsciousness for much of the time, and recovered from the injuries in a matter of weeks. But the Berger had bitten. This was going to be a hard one.

Things looked extremely promising, but there was one very large obstacle in the way of a follow-up trip; the group had only a hundred or so feet of ladder, and the shafts they had encountered so far had accounted for that length. Working into the small hours, trying to forget the factory and office jobs which demanded an early start each morning, the group constructed another 330 feet of ladder, using the old technique of soldering each rung in place. Even then they were dissatisfied; more ladder was needed if the next push was to be worthwhile. A Swiss caver was contacted and promised to turn up in mid-July with a plentiful supply of tackle from the main club in Geneva.

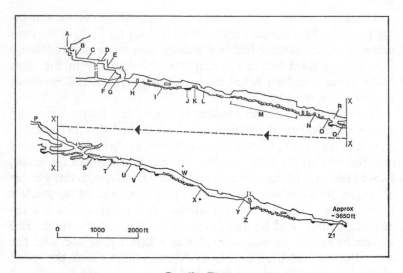

Gouffre Berger

A – Entrance
B – Cairn Hall
C – Winding Cleft
D – Boudoir
E – Garby's Shaft
F – Gontard's Shaft
G – Aldo's Shaft
H – Petzl's Gallery
I – Great Gallery
J – Lake Cadoux
K – Bourgin Hall
L – Little General's Cascade
M – Big Rubble Heap
N – Hall of the Thirteen

O – Balcony
P – Yves Inlet
Q – Pegasus Bridge
R – Germain Hall
S – Cloakroom
T – Abelle Cascade
U – Cascades
V – Claudine's Cascade
W – Grand Canyon
X – Gache's Shaft
Y – Little Monkey
Z – Hurricane Shaft
Z1 – Sump

11. A caver feeds lightweight electron ladder over a 150-foot pitch on a reconnaissance trip in Cave EK205, Mt Arthur, New Zealand. Note the use of a carbide light – essential on long exploratory trips with no chance of recharging electric light cells

12. A member of a British expedition to the Pierre St Martin, the world's deepest known cave, uses a dinghy to negotiate deep water between the Salle Monique and the Salle Susse

14th July. At the entrance to the Berger, conversation tailed off as carbide lamps were charged and lit and a final check was made of the tackle bags, bulging with ladder and rope. The Swiss caver had not yet turned up with his precious ladders, but his arrival after the main party had descended was greeted with even more than normal warmth by the French, who greedily eyed his rolls of ladder!

Ruiz Shaft, the Holiday Slide, the Winding Passage, the Boudoir – all were speedily negotiated until the previous obstacle was met again: Garby's Shaft. This proved to be 130 feet deep, and presented no particular difficulties. At the bottom, a small hole showed the way on – complete with a fierce and freezing draught; this made progress along the following winding rift a dry but very cold affair. After 260 feet of careful traversing along this rift, another pitch; here, Gontard belayed himself on to a tiny ledge to lifeline the others down. This shaft was only 85 feet, but led to a section of short drops and clambers in a wet and rather miserable passage. Things were to get a lot wetter.

Aldo Sillanoli volunteered for lifelining duties at the head of the next shaft, a usually thankless task (now rather superseded by the use of double lifelines left running through a belay at the top of the pitch, or self-lifelining devices) involving possibly many hours of waiting alone in the dark, inactive, lamp turned out to save fuel, sleep weighing heavily on eyelids. But this vigil was later rewarded by christening the place Aldo's Shaft; the same procedure gave Gontard's name a lasting place in the annals of caving. Neither, of course, knew this at the time, not that it would have been much consolation for the slow onset of shivering and cramp . . .

Georges Garby and Jean Cadoux climbed down the 130 feet of Aldo's Shaft, trying desperately but quite unsuccessfully to avoid the water which insisted upon accompanying them almost the entire way. They were now the only two left for the spearhead, all the others being left behind at various lifelining posts. They pressed on, into a small gallery. The walls and roof closed in as the sound of water thundering down the last shaft faded behind them. Still more narrowing bought nagging doubts into Cadoux's mind. Was their find to end like this?

Garby reached a small hole and peered through, whilst Cadoux watched his companion's head for the shaking which would mean the end. Through the tantalizing moments of silence, Cadoux heard something else, a deep rumbling. Could it be . . .? Garby spun round, his face beaming the answer as he moved to let Cadoux look too. Ahead lay an enormous chamber, and threading along the floor the Germe, the river they had dreamed about and sought for so many years. Their gushing full-blooded river link with the Vats of Sassenage somewhere in the vastness of rock beneath their feet.

Cadoux and Garby raced along the vast hall, drinking in the grandeur revealed by their barely adequate lights, until stopped by a deep lake. An attempt at traversing round it failed, so the two paused, soaking in the delicious sensation of being the first men ever to have seen this grand place, then headed for the surface. At each pitch, the lifeliner grinned delightedly at the news. At the foot of each pitch, the men waiting their turn to climb slumped to the floor in deep sleep, oblivious to the noise and cold. It was an expedition close to the limits of a 'one day' excursion: 27 hours.

On the next journey into the Berger, the French used abseiling techniques for the first time, and quickly appreciated how efficient this method of descent was. Cadoux did learn, however, that it is not wise to brake suddenly after a long abseil on nylon rope when very close to the bottom – he suffered three hard and undignified bounces on his rear end. The comedy was completed when the make-shift dam at the top of the shaft – Aldo's – was released when the dam controller interpreted the three splashes at the foot of the shaft as three footsteps. Barely had Cadoux recovered his wits from the puppet on a string act when a barrage of water descended from the heights! It is recorded that his comments were coarse.

Cadoux Lake was crossed without incident in a rubber dinghy, and the explorers were treated to the glorious and massive stalagmites of Bourgin Hall. Then a 30 foot pitch (the Little General's Cascade, called after the group's personal enemy at the time who had dyed the Berger stream with fluorescein and caused utter havoc in Sassenage when all the water turned bright green!) and a small drop of eight feet or so. Another few hundred feet brought them to another small pitch, but with no ladder to tackle it. Time had run out as well – or at least so they surmised, for each had relied on the others to bring a watch down, a reliance which turned out to have been misplaced. The point which this November attempt reached was 1,200 feet deep. Ahead lay a blackness so complete that the combined beams of the three cavers' lamps could not penetrate it.

The snows of that winter, as always, lay heavily upon the Sornin Plateau, effectively sealing off the cave entrances. Explorations were out of the question for months, but this was a good thing in one respect: it gave the group time to consider their next move. It was patently obvious that 'normal' caving trips were now out of the question; the November one had lasted 38 hours, with some of the lifelining parties having held their cold stations for 20 hours. A full-scale expedition was called for, with all its complex problems of trying to cater for the requirements of an unknown number of men (after the inevitable decimations of fatigue and illness) tackling an unknown number of subterranean obstacles. The basic strategy was settled, as a sort of compromise between some quite extreme suggestions: a base camp would be established at the lowest point possible, and a mobile advance camp would push ahead with the exploratory party.

The assault began on 14th July, 1954, with about a ton of equipment being jeeped, dragged or humped up the long hot slopes to the surface camp on the plateau. Try as hard as they might the group could not reduce the number of sacks to be taken underground, and each in his own way perfected a special series of oaths for every one. For the first time, the cavers were learning about large-scale underground work, the hard way. At times, when a particularly cumbersome sack caught for the umpteenth time in a constriction, they found themselves wondering, like all cavers in such circumstances, if it was really worth it. Like most cavers, they knew that it was – but then there was that enticing unexplored blackness beyond the 1,220-foot point.

Day after day the transporting of material continued, until at last 48 sacks were piled in Bourgin Hall – well over a quarter of a ton! A little farther, and the team stood at the very edge of unknown territory, and an annoying 15 foot drop. Here they rigged up a Tyrolienne traverse, a length of rope stretched tightly at a slight angle past the obstacle. Bags clipped to the rope then carried themselves down by the grace of gravity.

Now the Berger began to reveal some of its most precious treasures. The hall in which they found themselves was huge by any standards, 200 feet high by 150 wide in places, and suddenly there was the Hall of the Thirteen. Weariness was forgotten as they clambered among the enormous stalagmites and gour pools of this cathedral under the earth. It took a conscious effort to tear themselves away to search for a suitable camp area.

Continuing down the great chasm the next day, the French cavers came to the first real pitch of any length since Aldo's Shaft: a 60-foot climb from an overhanging platform which they called the Balcony. Dimly visible on the other side of this pitch was another stretch of passage continuing at the same level, but their way was obviously straight down. That was where the Vats lay. Doubts grew for a while at this point, though, for the team could hear a disturbingly large roar ahead of them. Obviously there was a massive waterfall to be passed, but did they have the necessary equipment? One man, impatient to know if this spelled the end of the expedition or not, dashed ahead. His loud guffaws were rather puzzling. Had he cracked under the strain of the enterprise? The others rounded a corner and came upon the 'Enormous Cascade' – a thin trickle of water falling into a calcite tube, the air compressed and escaping with the roar of a torrent through a small hole at the base!

The team pushed on through a lower wet part, ending in a series of small cascades; small, that is, but for the last one, Claudine's Cascade. Here the water plummeted straight down the ladder for the full 50 feet. Attempts to climb down came to nothing and the great trek back began. The expedition had lasted a week, gaining a total depth of 2,335 feet. Another expedition was agreed upon for the following summer, but the wait proved too much for the nucleus who paid a 'lightning' visit to the cave again in September. Their achievements on what turned out to be a 59-hour trip were quite remarkable. Not only did they reach Claudine's Cascade with their heavy load of steel scaffolding pipes which they used to hold the ladder clear of the waterfall, they added another 115 feet to the total depth.

Things were now hotting up considerably. The tackle had been left down the cave, and another push had to be made soon before winter closed in. It was made a fortnight later, towards the end of September, and brought the team triumph. An exhilarating and decidedly risky traverse at a steep angle down a gulf christened the

Grand Canyon, the floor 200 feet below the narrow sloping shelf along which the cavers climbed, led them to the 60 feet deep Gache's Shaft, and a world record of 2,962 feet. The following year saw a ten-day expedition push the depth of the apparently never-ending system by a 'mere' 270 feet, and it was not until 1956 that the French epic came to a close when three men managed to pass the 140 foot Hurricane Shaft to find the passage levelling off, then ending in what was described at the time as an impassable sump. The depth given then was 1,130 metres, 3,707 feet, but more recent survey work by British teams puts the depth of the sump at more like 3,650. A large discrepancy, true, but understandable to some extent with the extreme difficulties of underground surveying over such vast distances (over two and a quarter miles of horizontal distance). Whatever the precise figure is, 1956 spelled a world record for the Gouffre Berger team of French cavers, and their pleasure was well deserved.

Since 1956, numerous expeditions from other countries have visited the Berger, and some of the British ventures have been successful in terms of discoveries. In 1963 a British team managed to dive the first sump, which proved to be 240 feet long, but after only 60 feet of airspace came the second sump. Ken Pearce dived again in 1967 and managed to pass the long second sump, running out about 600 feet of line from base, to find another 300 feet of passage. He was alone, however, and had no tackle with which to descend the rift he found at the end. Pearce gained about 20 feet of depth. In 1968 French divers descended the rift which had stopped him and explored a short passage which ended in impenetrable sumps. Since then, no further progress has been made in forcing the connection (which many believe to be impossible) with the Vats of Sassenage.

Meanwhile, in the Pyrénées, roughly 350 miles wsw of the Gouffre Berger, is the Gouffre de la Pierre St Martin. The deepest known cave in the world, it straddles two countries, France and Spain, in its plummet to the valleys. The story of the piecing together of the component parts of the PSM – a handier label commonly used by the British, and which will be adopted here – is a complicated one, much more so than that of the Berger. It is necessary, then, to give only a summary.

The villagers of Sainte Engrâce – a tiny place lost in the mountain chain which divides France and Spain – grew used to the increasing numbers of cavers who roamed its valleys since the turn of the century. If they had any understanding of the motives of men who wanted to explore the open caves of the valley, this might understandably have been missing when cavers turned their attention to the mountains and plateaux above the village. But here, as at Sassenage, it was the top end of the system which cavers yearned to discover.

In 1950 two French cavers were wandering over the limestone plateau a few kilometres to the west of the Pic d'Anie. One of them was Georges Lépineux, a man possessing a healthy quota of native wit, and on this particular day the expression 'as the crow flies' took on a new meaning for him. When he noticed a crow flying up from the bottom of a nearby shakehole (crater in the limestone), his curiosity was aroused for he knew these birds only nested with an open space beneath them. He found a small opening and started pulling rocks away; a few fell into the hole, and Lépineux listened carefully for the tell-tale thump which would roughly indicate the depth beneath him.

No – this was impossible. He could not believe his ears, and threw down more stones. They fell, disrupting the black silence of countless centuries with their shrill howling whistle, for 1,100 feet. The hairs on the back of Lépineux' neck bristled; he was gaily gardening away at the top of the deepest natural shaft discovered at that time!

One year after the discovery, the French were ready to make their descent. In a shaft of this magnitude, electron ladders were obviously out of place, and abseiling/prusiking techniques had not developed to the stage where they could be considered, particularly in view of the boulder-covered ledges which jutted out in a few places. A winch was needed, and Max Cosyns, besides being a leading caver, dreamed winches. He designed one which resembled a bicycle – except instead of a front wheel it held a 440-yard drum of fine steel cable; foot and hand pedals provided the power. Whilst the whole contraption worked, so did the operators, pedalling the equivalent of 120 miles during each day of operation.

To honour his discovery – or did nobody else raise his hand? – Lépineux was lowered first. His journey of one and three quarters of an hour landed him on top of a slope of stone and boulders in a St Paul's Cathedral-sized chamber. Lépineux was hoisted back up, and Jacques Ertaud went down to take a few photographs but was so worried about not being able to connect up his body harness that he kept it on, though bent double in agony, during his short stay.

Marcel Loubens and Haroun Tazieff (the renowned ciné photographer of live volcanoes) descended next and set about exploration of the Salle Lépineux. They found a route out of the chamber into a huge new one, the Salle Elisabeth-Casteret, reaching a depth of 1,656 feet. Their bonus was the sound of a large river, audible beneath the boulder floor.

The 'Tour de France' winch had only just held out during the winching up of Loubens and Tazieff, so Cosyns designed a new electric model for the 1952 assault. Loubens went down first in that year, but Tazieff and two others could not follow until the following day due to a winch breakdown. Not much was said – there was little point – but the failure did not augur well.

After four days of exploring, Loubens decided to return to the surface to allow someone else to take his place. He struggled into his harness, clipped on to the swivel clamp on the end of the cable and telephoned the order to winch up. Slowly he started up the boulder slope, then the cable jerked and he stopped. The surface camp telephoned the dreaded report of another winch failure; Loubens hung there, no doubt thinking many deep thoughts about the collection of gears and ratchets on which his life would depend shortly.

After some hours the necessary repairs had been made, and the winch took the load. Slowly, Loubens began to rise up the shaft, after shouting a final goodbye to his companions who were to remain below. Then, with appalling suddenness, he fell. He landed on the slope, bouncing down for a hundred feet before coming to a halt.

What seems to make the disaster even worse in a sense is that Loubens did not fall from any great height in that giant of a shaft, a mere thirty or forty feet. A few minutes later and he would undoubtedly have been spared the agony of the 36 hours in which he took to die. Those who knelt by him through each bitter minute would have been spared much, too.

Two bolts should have been used to retain the clamp on the

swivel lock; there was only one. There was only one chance for Marcel Loubens.

As someone who regards with suspicion anything more mechanical than a brick, I have a profound admiration for Dr Mairey, the first man to entrust himself to the winch – and on a *descent* – after the accident. 'You must trust the clamp', a well-meaning friend told him just before he was lowered down. 'I don't', was all he said.

Equally brave were the five caving scouts from Lyons who immediately volunteered to ladder the shaft and steer the stretcher clear of obstructions. With perfect teamwork, they stationed themselves at depths of 263, 492, 699 and 787 feet where the worst of the potential snags were, but just as Louis Ballandraux had secured himself to the lowest of the ledges, the word was passed along. Loubens was dead.

His body was buried beneath a pile of stones, then exploration continued. The boulders at the end of Salle Elisabeth-Casteret resisted all attempts to pass through them; then Mairey and Tazieff noticed a hole in the wall above them and hastily scrambled up to it. A short passage led them to another chamber, not as high this time, but with gour pools and stalagmites covering the floor. The choice of name was obvious: the Salle Loubens. The roof lowered until it became more a tunnel than a chamber, and some way down this – the Métro – the two Frenchmen were delighted to discover the river of the PSM bursting out of the boulders covering the floor.

As far as cave potential was concerned, things were looking very good. They had reached the bed of impermeable shale, and the river should, by all accounts, continue at a steady angle all the way down to the valley resurgence. Less good was the exit from the cave. Tazieff had to hang from the end of the cable half-way up that enormous shaft for four and a half hours whilst frantic patch-up repairs were being done to the winch. Mairey had a different and probably even worse form of purgatory to go through: a night spent alone below, while the winch was stripped for better repairs, with Loubens's body, and the stench of blood filling the whole gloomy chamber.

A new much more solid winch was made for 1953, when one objective was to evaluate the possibility of recovering Loubens's body. French ace caver Norbert Casteret went down first, but could see no chance of fulfilling this objective with the equipment at the team's disposal. While Casteret and Mairey set off upstream a short way, the rest of the team set off down the Métro; an enormous roof collapse put paid to further progress that first day. A few days later a small team surmounted the obstacle, some fifty or sixty feet high, to continue into two more large chambers: the Salle Queffélec (no doubt named as an oblation to the designer and builder of that year's winch) and the Salle Adélie.

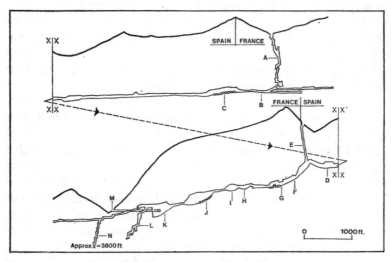

Gouffre de la Pierre St Martin

A – Tête Sauvage
B – Salle Cosyns
C – Salle Susse
D – Salle de Navarre
E – Lepineux Shaft
F – Salle Elisabeth
G – Salle Loubens

H – Salle Queffelec
I – Salle Adelie
J – Salle Chevalier
K – Salle de la Verna
L – Puits Maria Dolores
M – EDF Hydro-electric tunnel
N – Puits Parmant

In a do or die attempt, Lépineux and Daniel Epelly were winched down on 13th August with light camping equipment and food for three days. One by one they passed the chambers, discovered yet another, the Salle Chevalier, then stumbled across something which had them completely baffled. The river suddenly plummeted into open air! It was pitch black, and there were no stars; Lépineux and Epelly were flabbergasted. This did not make sense, they should still be in the depths of the earth. Then one thought to look at his watch – it was still daylight on the surface, so their 'open air' was in fact an enormous chamber. Their mistake was understandable, for the Salle Verna is 650 feet long, 400 feet wide and 330 feet no way on; the P S M was finished . . . or so it seemed.

They clambered down to the very bottom of the gigantic hall, to a record depth of nearly 2,400 feet, but to disappointment also for the water disappeared into the sand and gravel floor. There was no way on; the P S M was finished . . . or so it seemed.

The following year, plagued by access problems due to a border dispute between France and Spain (the 'Pierre St Martin' is actually a border marker between the two, putting the Lépineux Shaft in hot territory) the French launched another attack, but with the main purpose of recovering Loubens's body.

On paper this task does not sound particularly daunting. Given an efficient winch and a stout container, surely a 1,100 foot shaft could be tackled in eleven times the time taken on a 100 foot shaft? In practice, the enterprise took on epic proportions, with a high degree of risk to all concerned. The hoisting of the aluminium rocket-shaped coffin took 20 hours, and for 13 of those José Bidegain matched the coffin's ascent inch by inch by using a hand-powered autohoist. Time and time again he pushed the coffin, clanging like a cracked bell, away from some obstruction, constantly in fear of the heavy winch cable cutting through his own lighter one. Eventually, Loubens was raised from over a thousand feet down and reburied in a shallower if more permanent grave.

Although the cavers had been categorically forbidden to do any exploratory work on this occasion, they felt justifiably annoyed at this intrusion of political red tape. After all, permission had been given to raise Loubens's body; surely it could be realized that Loubens himself would have been bitterly disappointed that officialdom could stop the cavers where no natural obstacles had, nor his own death.

Quietly, a small party made its way upstream, passing the point previously reached and arriving, after one kilometre of passage, at the long Salle de Navarre. Beyond lay a low tunnel, its walls curling in over very deep water, and out of this tunnel – Tunnel du Vent – shrieked an icy wind. They could go no farther.

The next phase in the exploration of the PSM came when engineers blasted a tunnel into the Salle de la Verna itself in the first stage of a giant hydro-electric scheme. The scheme failed, but cavers were left with a very rapid means of entry to the bottom of the cave, and had the added bonus of the discovery of the Gouffre d'Arphidia half-way along the tunnel.

By now, work in the PSM was on an international scale, and three Spanish cavers were convinced that such a system could not simply dribble into nothing at the Verna. They reasoned that if one could find the continuation of the shale bed on the far wall of the Verna – through which bed the river had, in ages past, cut through to tear out the huge chamber below – they would find, above it, the original passage. Two of them, Felix de Arcaute and Juan de San Martin, made a hair-raising ascent of the great and crumbling Verna wall. At 300-plus feet, San Martin gave the wall a vicious kick in frustration – to find a neat hole suddenly appear round his boot. The two Spaniards scrabbled away at the hole until they could get their bodies through into the 1,000 foot long Galerie Aranzadi. That was in 1961, and since then much effort has been poured through that hole in the Verna wall, extending it considerably horizontally and vertically, too, with the Reseau Maria-Dolores and even deeper Puits Aziza. The hope has always been that a shaft from this downstream section might connect once again with the Pierre's river, but this has not been realized.

In 1964, the gruelling fight upstream was recommenced by the Spéléo-Club de Paris who pushed past the farthest point reached by the Spanish in 1960; another expedition the following year added another kilometre upstream leading to the 650 foot long chamber, the Salle Susse.

By 1966, two important developments were under way, French, Spanish and Belgian cavers had founded the Association des Recherches Spéléologiques Internationales de la Pierre St Martin (ARSIP) with the dual purpose of co-ordinating exploration of the system and restricting access to recognized groups prepared to join in with the association's plans. And careful searches were being made of the Arres d'Anie, the huge limestone plateau directly above the Salle Susse. One by one, the thousands of openings on the plateau were descended, but particular attention was paid to one, a 236-foot shaft known simply as D9, which was blocked by boulders at the bottom.

Climbing back up D9, one sharp-eyed lad noticed a hole in the wall and promptly insinuated himself into it, feet first. The angle steepened to about 45° so progress was easy – he simply slid down. A willy-nilly method of progress such as this is not the most commendable in virgin territory, as was soon made clear when he found his legs dangling over a pitch. Frantic braking with hands, arms and bottom brought him to a halt just in time! The drop below him was almost 250 feet clear.

Every scrap of available ladder was carted over the plateau to D9, and the descent was made over a number of days. At the bottom of the Tête Sauvage, as D9 is now known, the explorers found a large river; a little way downstream they spotted a piece of paper on a ledge above the water. Written on it was 'Fin de topo [end of survey] 1966'. They were in the Pierre St Martin, and in that moment of connection had increased its depth to no less than 3,780 feet. Having been passed previously by the Gouffre Berger, the PSM was once again the deepest cave in the world.

The situation in and around the PSM is a complicated one at the moment, but slowly more pieces are being added to the jigsaw and it seems only a matter of time before the PSM is surpassed by one of its immediate neighbours. Caving teams from many countries besides France and Spain gather in the area each summer, and expeditions there figure as a regular feature in three or four British club programmes. The British in particular have high hopes of being able to push the bottom end of the PSM itself, via the point which marks the deepest part of the cave.

In 1970, a parallel system to the PSM was found, the Gouffre Lonné-Peyrét, and this was explored for many miles to a final depth of 2,355 feet in the Salle Stix. The really intriguing thing is that the Salle Stix was found by careful surveying to be only a hundred feet away from a part of the Gouffre d'Arphidia, the system discovered by chance when the hydro-electric tunnel was blasted into the Verna.

The outline of the jigsaw is plain; only the pieces have to be fitted in. Another entrance must be found into the Galerie Wessex – the upstream end of the Gouffre Lonné-Peyrét – adding some 6-700 feet to the cave. Then some link must be forged over that tantalizingly short distance between the bottom of the Lonné-Peyrét, 2,352 feet deep, and the Arphidia, which would add another 1,377 feet. And when the sum of $1+1+1=4,429$ feet, any caver worth his salt will do his utmost in making those additions. A cave, though, can be a hard teacher, and the Lonné-Peyrét delivered a particularly savage lesson in 1971. A French caver, leading a small team upstream, became exhausted while climbing an 80-foot wet pitch. As he was using a self-lifelining device, he found himself locked on the rope and too weak to free himself. His companions were unable to help him, and after two hours under the waterfall he died from drowning and exposure.

Perhaps the Lonné-Peyrét will be linked with the Arphidia; perhaps the French will find the long-sought Rivière St Georges system, with its potential even greater than the adjacent PSM; perhaps M3 pot will 'go', providing an even higher entrance to the PSM than the Tête Sauvage. In caving, 'perhaps' has to move over for actuality, however, and the Pierre St Martin remains not just deep, but the deepest.

Chapter 13

Epilogue

On quite a few occasions, I have drawn parallels between mountaineering and caving; yet there is one huge difference, a quite fundamental division. A mountain is a roughly pyramidal lump of rock, present in reality right up to the topmost needle point. A cave is just the opposite: an absence of rock to the extent that a man can actually move inside it. Probably it is as well that cavers are usually blessed with a bountiful, if sometimes rather warped, sense of humour; they certainly need it in order to spend weekend after weekend, holiday after holiday, risking their necks to find out just how far this absence extends. The caver displays a perfect love–hate relationship with rock; he needs it to provide the walls of his cave (although it's a nice idea, caving in the open), yet he curses it to hell and back when it pinches in slightly, marking an end to the hole.

Fortunately for caving, the world is generous in its outlay of good thick limestone deposits, courtesy of bygone oceans. Less fortunately, caves have a habit of being shy about betraying their whereabouts. In terms of large-scale expeditions overseas, there are still plenty of unexplored holes waiting, quite open to the outside world. But with expeditions abroad being organized each year by almost every major caving club, and quite a few minor ones too, it will not be that many years before all the plums have gone, and visiting cavers will have to devote much more time to digging, blasting and diving in order to reach their objectives.

In the meantime, expeditions are enjoying something of a Golden Era, similar to that phase in mountaineering when attention wandered farther afield than the Alps to the Himalayas, the Karakoram, and the Andes. For some reason, the really big overseas exploits are being carried out mainly by British expeditions; perhaps the apparent lack of interest on the part of other

Caving regions of the world

European countries is because they still have plenty on their own doorsteps!

The first two serious caving expeditions to the Himalayas both took place in the same year, 1970, and both were British. When I discussed the prospects of the British Karst Research Expedition with leader Tony Waltham, he expressed delightfully reserved optimism. After all, as he pointed out, there is great theoretical potential in the area, with the limestone around Annapurna reaching a thickness of 12,000 feet, and meltwater streams falling down through 11,000 feet of this. But nobody had done any work there before, at any reasonable altitude, and virtually the only literature he could refer to was Herzog's book on Annapurna, in which there is a casual reference to some caves high up. It was not much to go on, but it provided the spark which such ventures need, the moment of conception when the caver is lurched out of his lethargy and finds his imagination racing off into a new dimension. Planning follows imagination, and you have an expedition.

Waltham's reservations were justified, for his team found little of significance at altitude, but did explore nearly a mile of the low-lying Harpan River Cave. They were of the opinion that the absence of caves in the Nilgiri Limestone – their prime target – was due mainly to the unjointed nature of the rock. The lack of this important feature in cave formation was a local one, and certainly does not preclude possibilities for deep systems in the other great ranges. The other enterprise in the field that year, the British Speleological Expedition to the Himalayas, reached its base camp – a field commandeered for their sole use by the Prince of Arki – only by frequent tyre lowering at low bridges on 'Arry, a mammoth Preston double-decker bus! Despite their auspicious mode of transport, this expedition had little of encouragement to report. Still, the Himalayas are big hills, matched by the enthusiasm and dogged perseverance of that band of cavers who do not have the Pierre St Martin, or the Hölloch, or the Flint Ridge-Mammoth Cave on their doorstep.

Whilst the Himalayas might have been expected to contain a surprise or two – a four or five thousand foot pot would have been nice, just to whet the appetite – not even the most far-sighted of pundits expressed much hope for surprises from the Speleological Reconnaissance Expedition to Iran in 1971. Much digging out of geological surveys revealed fat wedges of limestone in the Zagros mountain chain, but Iran just did not *sound* like the right kind of place. Whilst later developments never cure us of our 'sixth sense', thank God we have grown cautious enough never to rely too heavily upon it.

The expedition pinpointed the northern end of the Zagros chain as being most useful; this was where the rainfall was highest – an odd prerequisite for a Britisher going abroad, but a vital one when he is looking for holes.

Most of the time was spent prospecting on and around the mountain Kuh-e-Parau (11,131 feet), a blunt lump of rock which I recall seeing, through sweat-stung eyes, when travelling through the nearby town of Kermanshah in the early 'sixties. Even had had I been aware of the many shakeholes on the mountain's high plateau, the heat would certainly have deterred me from making the long climb up – and it was the heat, plus the half imagined, half real fear of wandering leopards, which proved the 1971 team's greatest problem.

For various reasons the expedition did not get around to examining Ghar Parau, the most obvious entrance, until quite late in the proceedings. By this time enthusiasm had waned a little, and no great hopes were held out; at best, they expected a pleasant ramble down a cave for perhaps a couple of hundred feet. Events became a little embarrassing, though, when Ghar Parau refused to stop, and burrowed its way into the mountain for hundreds upon hundreds of feet. Frantically pouring every ounce of effort into an eleventh-hour descent, some of the team reached a final depth of 2,300 feet – and the cave was still going, down a pitch and on into the blackness.

Slightly shell shocked, the cavers returned to Britain and immediately began planning for a return visit. In view of their remarkable discovery, and clear prospects of a record breaker, financial and material backing for the second attempt was plentiful – more so, probably, than for any previous caving expedition. So rosy was the financial side, in fact, than on their return and when all obligations had been covered, the team was able to set up a Ghar Parau Foundation – an enterprise similar in function to the Mount Everest Foundation, but helping cavers rather than climbers.

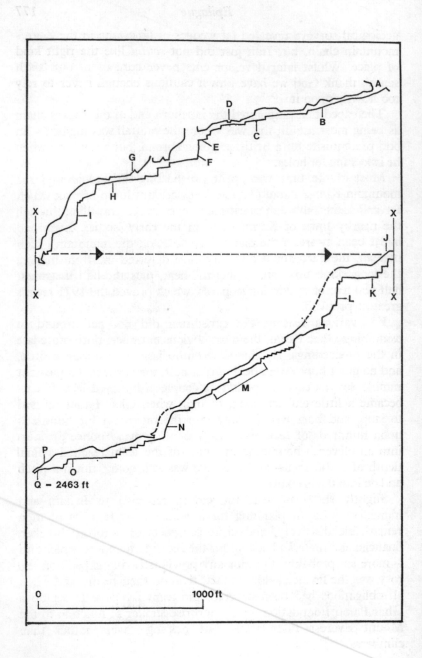

Led by Dave Judson, one of Britain's most competent and best-known cavers, the 1972 expedition descended upon Ghar Parau with the best of manpower, materials, techniques and intentions. The push downwards was a wearying one, though, for unlike most other systems, Ghar Parau obstinately grows more and more difficult the deeper one penetrates. What did it matter – dreams were filled with the glories which would surely lie below and beyond the twenty-sixth pitch.

The twenty-sixth pitch was reached, rigged, climbed. Virgin territory. Around a corner, walls lined with clinging mud. Would the next pitch be a giant, or would Ghar Parau condescend to let the explorers reach their record depths by way of giant halls gently inclining for mile after mile? There was no next pitch; there were no miles of passage; just a small, foul sump. An inglorious full-stop, an end to the dreaming.

In the Southern Hemisphere, New Zealand and Australian cavers have developed their techniques to a high pitch in recent years. Much of the fresh enthusiasm has been injected by immigrant Englishmen seeking rather more cave potential than could be found in their home country; the same thing has been witnessed in Canadian caving also. Single rope techniques of abseiling and prusiking have been widely adopted in Australasia, enabling cavers to really get to grips with such tough pots as Tasmania's 1,054-foot Khazad-dum (ex-Tolkien names are coming thick and fast down-under!). Such are the advantages of reduced tackle weight and bulk by these methods that two cavers, Peter Shaw and Philip Robinson, were recently able to bottom 758-foot deep Tassy Pot as a duo. A far cry indeed from the big teams lumping leviathan rope ladders and hemp lifelines in the sport's earliest era.

Ghar Parau, Iran

A – Entrance	J – First Traverse
B – Greasy Slab	K – Slid Traverse
C – Damavand Avenue	L – Shahanshah
D – Corkscrew	M – The 39 Steps
E – Cyrus the Great	N – Yek Do Se Pitch
F – The Eroica	O – High Way Grotto
G – Masjid Hall	P – Ferdowsi Avenue
H – Quality Street	Q – Sump
I – Bootlace Pitch	

Whilst European cavers fret over the prospects of further footage in the Pyrénées, an Australasian expedition is even at the moment of writing hacking its way through the jungles of the Southern Highlands in Papua New Guinea. There lies a belt of limestone over five thousand feet thick, bombarded by an annual rainfall of between one hundred and two hundred inches.

And there stands the huge Hindenburg Wall . . . 'The Wall presents a spectacle of limestone formations that boggle the mind. Flying west one sees an area so remote, so isolated and weird that it is hard to believe you are still on the same planet. Far below you is spread a large plateau covered by stunted, moss-enshrouded vegetation; large blocks hundreds of feet high rise above the plateau in weird shapes, often with silver rivers of water cascading out of them; ragged cliffs point dramatically at the sky, deep ravines threaten to take your breath away . . . The south side drops 8,000 feet to the plains of the Fly River. We found the only sensible way to explore the Hindenburg Wall was by helicopter. We were often amazed by huge rivers that poured out of the wall, often exiting half-way up and cascading thousands of feet into the jungle below. Into one of these entrances we were able to fly for about 200 feet.' So wrote Malcolm Robb describing the initial reconnaissances of the area, and one cannot help being impressed by that last sentence!

Surely, deeper systems than the Pierre St Martin will be found in the Northern Hemisphere, but who knows what exciting possibilities lie waiting in the jungles of Papua New Guinea? One thing is certain: the Australians and New Zealanders will not rest until they have found out, and, as I write, a group of twenty cavers in Britain are planning the most ambitious, and probably most costly, caving expedition ever, to that region.

Caving is all this. Caving is Harry Pearman's, 'Because it *might* be there'. Caving is the cold, the dark, the wet, the fear, the back-breaking thousand foot crawl, the long lonely pitch on slender ladder, the purposelessness of it all . . . until . . . a tiny pocket in the rock guarding one exquisite formation, a cushion of sparkling crystals. For the moment, at that place, your light is the only light in the world – your light is the world.

Caving is all this, in . . . The silence. The blackness.

A Short Glossary of Caving Terms

ABSEIL: A method of sliding down a single or double rope in a controlled manner.

ACTIVE CAVE: One which still takes water. A dry cave is called dead, or inactive.

ADIT: A nearly level entrance to a mine, often used for drainage.

AVEN: A vertical hole in the roof of a passage, sometimes leading to higher passages.

BED: One of the strata in sedimentary rocks.

BEDDING PLANE: The plane which separates two layers of rock; in caving, usually a low, wide passage.

BELAY: Secure natural or artificial anchorage for ropes and ladders; the attachment of such equipment to the anchorage; the wire or other ropes used for such anchorages; the safeguarding of a climbing caver by a lifeliner.

BRIDGE: An arch of rock spanning from one cave wall to another.

CALCITE: Crystalline form of calcium carbonate, $CaCO_3$; the basic constituent of cave formations.

CARBIDE: Calcium carbide, CaC_2; reacts with water in a carbide lamp to give acetylene gas for lighting.

CAVE: A natural subterranean cavity in rock. From the exploratory point of view, a cave – as opposed to a pothole – is developed more or less horizontally and usually requires no ladders or ropes in its exploration.

CAVE PEARL: A more or less spherical calcite formation formed in an active cave pool.

CHAMBER: A large cavity in a cave system.

CHIMNEY: A narrow vertical fissure which can be climbed by chimneying technique, i.e. the back and hands pressed against one wall, the feet against the other.

CHOKE: A total or partial blockage of a passage by boulders (then known as a boulder choke), silt or clay.

CLAUSTROPHOBIA: A morbid fear of confined spaces.

CLINTS: Network of isolated blocks on an exposed limestone plateau, caused by corrosion of vertical joints.

COLUMN: The formation resulting from the joining of a stalactite to a stalagmite.

CORRASION: Abrasion of a channel by particles of rock suspended in water.

CORROSION: Chemical erosion of rock by slightly acid water.

CRAWL: Section of low passage demanding progress on hands and knees or on stomach.

CURTAIN: Flat, thin formation hanging down from cave roof or wall, often wavy.

DESCENDEUR: A mechanical device used for abseiling. There are many different types.

DIG: A site of excavation, underground or on the surface, dug or blasted by the caver in an attempt to reach new passages or a new cave.

DIP: The angle (in degrees) of the slope of a rock stratum.

DOLINE: See SHAKEHOLE.

DRAG SHEET: A stretcher for cave rescue in constricted places, formed from a sheet of canvas or rubber.

DRIPSTONE: General term for formations formed by falling drops of water.

DRY SUIT: Waterproof suit made from one or more layers of rubber or rubberized canvas.

DUCK: The term generally used nowadays for a point where the cave roof very nearly meets the water surface; requires partial or virtually complete immersion of the head to pass.

ELECTRON LADDER: Modern form of caving ladder, superseding the old rope ladder; light aluminium alloy rungs attached to galvanized cable sides.

EROSION: Removal of rock by chemical and/or physical action.

EXPANSION BOLT: A metal bolt used for artificial belays; inserted into a hole drilled in the rock, the bolt grips when hammered in, or when a protruding nut is tightened.

EXPOSURE: The dangerous lowering of a caver's core or trunk temperature by the effects of cold, wet, and inadequate food intake; maybe fatal if not treated promptly.

EXPOSURE SUIT: Suit to protect against the effects of cold; see DRY SUIT, GOON SUIT, WET SUIT.

FALSE FLOOR: Layer of horizontal calcite stretching out from the wall or walls of a cave passage; formed when a coating of calcite is left suspended by the washing away of the supporting base of cave sediment and boulders.

FAULT: A fracture in the rock resulting in the strata on one side being discontinuous with those on the other side.

FLOWSTONE: Covering of calcite on a cave wall or floor.

FLUORESCEIN: An extremely powerful, non-toxic dye used for tracing the passage of underground watercourses.

FORMATION: Any one of the many forms of cave decoration formed by the deposition of calcium carbonate, frequently coloured by traces of minerals; also crystal decorations, chiefly gypsum.

GOON SUIT: A type of dry suit developed for survival at sea.

GOUR POOL: A pool created by the formation of a curved wall of calcite deposited by water flowing over its lip.

GROUND WATER: Water below the water table.

GYPSUM: Hydrated calcium sulphate ($CaSO_4.2H_2O$); responsible for some particularly beautiful cave decorations.

HEADER: Cable harness for connecting the top of an electron ladder to a belay wire or rope.

HELICTITE: Usual term for the stalactitic formation which forms apparently contrary to gravity, its often extremely delicate extensions twisting in a seemingly random manner.

HISTOPLASMOSIS: A disease, chiefly confined to certain parts of the Americas and South Africa, which may be caught by the inhalation of cave dust where there are infected bat, bird or other animal droppings.

HYDROLOGY: The study of subterranean water movement.

HYPOTHERMIA: See EXPOSURE.

JOINT: A vertical division in a stratum of rock through to the bedding plane; one crossing several strata is known as a master joint.

KARABINER: A metal snap-link, generally oval in shape, with a spring-loaded gate on one side to allow it to be clipped to a rope, ladder, belay, etc; colloquially known as a KRAB.

KARST: A region in Yugoslavia which has given its name to typical exposed limestone terrain.

LIFELINE: The safety rope, controlled by the lifeliner, used to safeguard a caver on a ladder or dangerous climb.

LIMESTONE: Rock which contains at least half, by weight, of calcium carbonate.

MASTER-CAVE: The main drainage channel of a cave system or systems.

MAYPOLE: Sections of metal tube joined together inside the cave to raise a ladder to a high-level passage entrance.

MEANDER: Semi-circular bend in a passage.

NEIL ROBERTSON STRETCHER: A stretcher developed for rescue from confined hatchways on ships, now modified for cave rescue.

NEOPRENE: A synthetic rubber, the foam version of which is used in the manufacture of wet suits.

OGOF: Welsh word for cave.

OXBOW: A loop of passage abandoned by the stream.

PHREATIC ZONE: Saturated zone of permeable rock; a phreatic cave is one formed by water corroding rock below the water table.

PITCH: Vertical descent in a cave, usually requiring ladder or at least a handline.

PITON: A metal spike which can be driven into natural rock fissures to give a belay point.

POTHOLE: Generally, the term for a cave system in which there are shafts which require the use of tackle for descent.

PRUSIKING: A technique for ascending a fixed rope using prusik or other knots, or mechanical prusik devices; the knot or device will slide upwards when pushed, but locks on the rope when a pull is applied.

RAPPEL: See ABSEIL.

RESURGENCE/RISING: The point at which an underground stream emerges on the surface.

RIMSTONE POOL: See GOUR POOL.

RUCKLE: Generally, the Mendip caving term for a boulder choke.

SCALLOP: Oval, concave markings on the walls of a cave passage, indicating the direction of flow of the current which made them.

SHAKEHOLE: A depression (doline) in a limestone region indicating the possible presence of a cave entrance, or formed by the collapse of a cave roof.

SINGLE ROPE TECHNIQUES: One term covering both abseiling and prusiking techniques combined in the exploration of a cave.

SINK: See SHAKEHOLE.

SIPHON: Loosely, a sump. Less commonly used nowadays.

SKYHOOK: A device for raising and securing an electron ladder to an expansion bolt previously fitted in some high, inaccessible spot; obviates the need for leaving a ladder underground permanently.

SLOCKER: Mendip term for cave or shakehole.

SPELEOLOGY/SPELAEOLOGY: Caving from a strictly scientific point of view, hence SPELEOLOGIST.

SPELUNKER: An American term for caver.

SQUEEZE: Constriction in a cave passage so tight that the caver has to physically force his body through.

STALACTITE: A cave formation hanging straight down from the roof, usually of calcite.

STALAGMITE: A cave formation growing straight up from the floor, often with a corresponding stalactite directly above.

STRAW : A thin, hollow stalactite.

SUMP : That point where the cave roof dips below the water level; sometimes passable by diving.

SWALLET : Openings in limestone taking a stream or streams.

TRAVERSE : A more or less horizontal climb along ledges or holds at a high level above a chamber or passage; to proceed along such a climb.

VADOSE ZONE : The zone above the water table.

WATER TABLE : The level below which all permeable rock and cave passages are saturated; marks the top of the phreatic zone.

WET SUIT: A protective one- or two-piece suit made from neoprene foam rubber, sometimes lined with stretch nylon fabric. Water absorbed into the foam cells is kept warm by the caver's body heat. A form of EXPOSURE SUIT.

Index